Chapter 3 Mechanics 1

Chapter 4 Statistics 1

Chapter 5 Decision Mathematics 1

Specification information

Edexcel Mathematics

MODULE	SPECIFICATION TOPIC	CHAPTER REFERENCE	STUDIED IN CLASS	REVISED	PRACTICE QUESTIONS
Core 1	Algebra and functions	1.1			
	Coordinate geometry in the (x, y) plane	1.2			
	Sequences and series	1.3			
	Differentiation	1.4			
	Integration	1.5			
Core 2	Algebra and functions	2.1			
	Coordinate geometry in the (x, y) plane	2.2			
	Sequences and series	2.3			
	Trigonometry	2.4			
	Exponentials and logarithms	2.5			
	Differentiation	2.6			
	Integration	2.7			
Mechanics 1	Vectors in mechanics	3.1			
	Kinematics in a straight line	3.2			
	Statics of a particle	3.3			
	Moments	3.3			
	Dynamics in a straight line or plane	3.4			
Statistics 1	Representation and summary of data	4.1			
	Probability	4.2			
	Correlation and regression	4.5			
	Discrete random variables	4.3			
	The normal distribution	4.4			
Decision 1	Algorithms	5.1			
	Algorithms on graphs	5.2			
	The route inspection problem	5.2			
	Critical path analysis	5.3			
	Linear programming	5.4			
	Matchings	5.5			

Examination analysis

The assessment is by written papers. All questions are compulsory.
Core 1 + Core 2 + one of Mechanics 1, Statistics 1 and Decision 1.

Core 1	AS	No calculator	1 hr 30 min exam	33.3%
Core 2	AS	Scientific/graphic calculator	1 hr 30 min exam	33.3%
Mechanics 1	AS	Scientific/graphic calculator	1 hr 30 min exam	33.3%
Statistics 1	AS	Scientific/graphic calculator	1 hr 30 min exam	33.3%
Decision 1	AS	Scientific/graphic calculator	1 hr 30 min exam	33.3%

Letts

Revise AS

Edexcel Mathematics

Contents

Your AS/A2 Level Mathematics course

AS and A2

The Edexcel Mathematics A Level course is made up of six modules. Students first study the AS (Advanced Subsidiary) course, which has three modules. Some will then go on to study the second part of the A Level course, a further three modules, called A2. Advanced Subsidiary is assessed at the standard expected halfway through an A Level course: i.e., between GCSE and Advanced GCE. This means that AS and A2 courses are designed so that difficulty steadily increases:

- AS Mathematics builds from GCSE Mathematics
- A2 Mathematics builds from AS Mathematics.

How will you be tested?

Assessment units

For AS Mathematics, you will be tested by three assessment units. For the full A Level in Mathematics, you will take a further three units. AS Mathematics forms 50% of the assessment weighting for the full A Level.

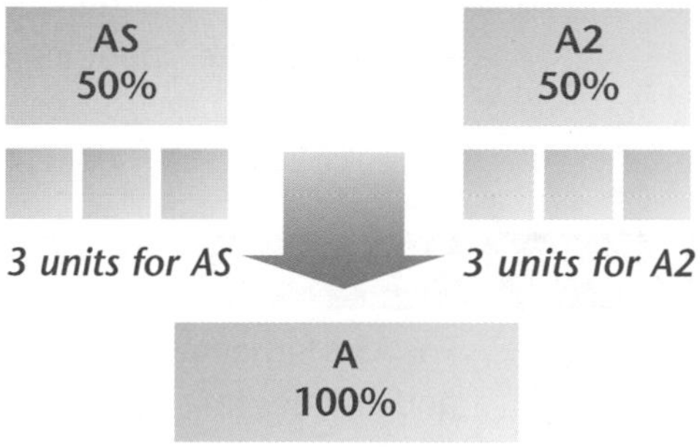

Each unit can normally be taken in either January or June. Alternatively, you can study the whole course before taking any of the unit tests. There is a lot of flexibility about when exams can be taken and the diagram below shows just some of the ways that the assessment units may be taken for AS and A Level Mathematics.

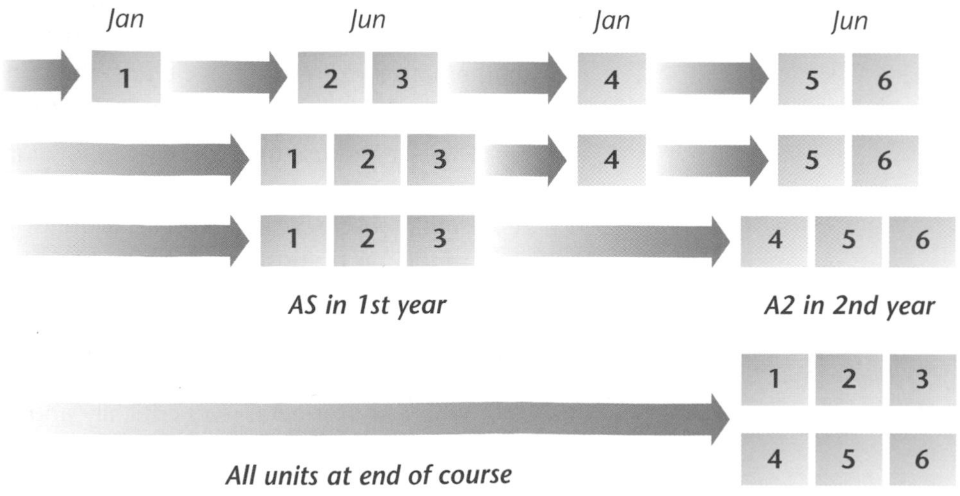

If you are disappointed with a module result, you can resit the module. There is no restriction on the number of times a module may be attempted. The best available result for each module will count towards the final grade.

Synoptic assessment

The GCE Advanced Subsidiary and Advanced Level Qualification specific Criteria state that A Level specifications must include synoptic assessment (representing at least 20% of the total A Level marks).

Synoptic assessment in mathematics addresses candidates' understanding of the connections between different elements of the subject. It involves the explicit drawing together of knowledge, understanding and skills learned in different parts of the A Level course, through using and applying methods developed at earlier stages of the course to solving problems. Making and understanding connections in this way is intrinsic to learning mathematics.

Key skills

It is important that you develop your key skills throughout your AS and A2 courses. These are important skills that you need whatever you do beyond AS and A Levels. To gain the key skills qualification, which is equivalent to an AS Level, you will need to collect evidence together in a 'portfolio' to show that you have attained Level 3 in Communication, Application of number and Information technology. You will also need to take a formal test in each key skill. You will have many opportunities during AS Mathematics to develop your key skills.

It is a worthwhile qualification, as it demonstrates your ability to put your ideas across to other people, collect data and use up-to-date technology in your work.

What skills will I need?

For AS Mathematics, you will be tested by assessment objectives: these are the skills and abilities that you should have acquired by studying the course. The assessment objectives for AS Mathematics are shown below.

Candidates should be able to:

- recall, select and use their knowledge of mathematical facts, concepts and techniques in a variety of contexts
- construct rigorous mathematical arguments and proofs through use of precise statements, logical deduction and inference and by the manipulation of mathematical expressions, including the construction of extended arguments for handling substantial problems presented in unstructured form
- recall, select and use their knowledge of standard mathematical models to represent situations in the real world; recognise and understand given representations involving standard models; present and interpret results from such models in terms of the original situation, including discussion of the assumptions made and refinement of such models
- comprehend translations of common realistic contexts into Mathematics; use the results of calculations to make predictions, or comment on the context; and, where appropriate, read critically and comprehend longer mathematical arguments or examples of applications

- use contemporary calculator technology and other permitted resources (such as formulae booklets or statistical tables*) accurately and efficiently; understand when not to use such technology, and its limitations; give answers to appropriate accuracy.

  * A copy of the formulae booklet and statistical tables can be found on the A Level mathematics section of the Edexcel website, www.edexcel.com

Progression and prior learning

Mathematics is, inherently, a sequential subject. There is a progression of material through all levels at which the subject is studied. The criteria therefore build on the knowledge, understanding and skills established at GCSE.

Thus, candidates embarking on AS/A2 Level study in Mathematics subjects are expected to have covered all the material in the Higher Tier. This material is regarded as assumed background knowledge. However, it may be assessed within questions focused on other material.

Exam technique

What are the examiners looking for?

Examiners use certain words in their instructions to let you know what they are expecting in your answer. Make sure that you know what they mean so that you can give the right response.

Write down, state

You can write your answer without having to show how it was obtained. There is nothing to prevent you doing some working if it helps you, but if you are doing a lot then you might have missed the point.

Calculate, find, determine, show, solve

Make sure that you show enough working to justify the final answer or conclusion. Marks will be available for showing a correct method.

Deduce, hence

This means that you are expected to use the given result to establish something new. You must show all of the steps in your working.

Draw

This is used to tell you to plot an accurate graph using graph paper. Take note of any instructions about the scale that must be used. You may need to read values from your graph.

Sketch

If the instruction is to sketch a graph then you don't need to plot the points but you will be expected to show its general shape and its relationship with the axes. Indicate the positions of any turning points and take particular care with any asymptotes.

Find the exact value

This instruction is usually given when the final answer involves an irrational value such as a logarithm, π or a surd. You will need to demonstrate that you can manipulate these quantities so don't just key everything into your calculator or you will lose marks.

If a question requires the final answer to be given to a specific level of accuracy then make sure that you do this or you might needlessly lose marks.

Some dos and don'ts

Dos

Do read the question

- Make sure that you are clear about what you are expected to do. Look for some structure in the question that may help you take the right approach.
- Read the question *again* after you have answered it as a quick check that your answer is in the expected form.

Do use diagrams

- In some questions, particularly in mechanics, a clearly labelled diagram is essential. Use a diagram whenever it may help you understand or represent the problem that you are trying to solve.

Do take care with notation

- Write clearly and use the notation accurately. Use brackets when they are required.
- Even if your final answer is wrong, you may earn some marks for a correct expression in your working.

Do avoid silly answers

- Check that your final answer is sensible within the context of the question.

Do make good use of time

- Choose the order in which you answer the questions carefully. Do the ones you find easiest first.
- Set yourself a time limit for a question depending on the number of marks available.
- Be prepared to leave a difficult part of a question and return to it later if there is time.
- Towards the end of the exam make sure that you pick up all of the easy marks in any questions that you haven't got time to answer fully.

Don'ts

Don't work with rounded values

- There may be several stages in a solution that produce numerical values. Rounding errors from earlier stages may distort your final answer. One way to avoid this is to make use of your calculator memories to store values that you will need again.

Don't cross out work that may be partly correct

- It's tempting to cross out something that hasn't worked out as it should. Avoid this unless you have time to replace it with something better.

Don't write out the question

- This wastes time. The marks are for your solution!

What grade do you want?

Everyone should be able to improve their grades but you will only manage this with a lot of hard work and determination. The details given below describe a level of performance typical of candidates achieving grades A, C or E. You should find it useful to read and compare the expectations for the different levels and to give some thought to the areas where you need to improve most.

Grade A candidates

- Recall or recognise almost all the mathematical facts, concepts and techniques that are needed, and select appropriate ones to use in a variety of contexts.
- Manipulate mathematical expressions and use graphs, sketches and diagrams, all with high accuracy and skill.
- Use mathematical language correctly and proceed logically and rigorously through extended arguments or proofs.
- When confronted with unstructured problems they can often devise and implement an effective solution strategy.
- If errors are made in their calculations or logic, these are sometimes noticed and corrected.
- Recall or recognise almost all the standard models that are needed, and select appropriate ones to represent a wide variety of situations in the real world.
- Correctly refer results from calculations using the model to the original situation; they give sensible interpretations of their results in the context of the original realistic situation.
- Make intelligent comments on the modelling assumptions and possible refinements to the model.
- Comprehend or understand the meaning of almost all translations into mathematics of common realistic contexts.
- Correctly refer the results of calculations back to given context and usually make sensible comments or predictions.
- Can distil the essential mathematical information from extended pieces of prose having mathematical content.
- Comment meaningfully on the mathematical information.
- Make appropriate and efficient use of contemporary calculator technology and other permitted resources, and are aware of any limitations to their use.
- Present results to an appropriate degree of accuracy.

Grade C candidates

- Recall or recognise most of the mathematical facts, concepts and techniques that are needed, and usually select appropriate ones to use in a variety of contexts.
- Manipulate mathematical expressions and use graphs, sketches and diagrams, all with a reasonable level of accuracy and skill.
- Use mathematical language with some skill and sometimes proceed logically through extended arguments or proofs.
- When confronted with unstructured problems they sometimes devise and implement an effective and efficient solution strategy.
- Occasionally notice and correct errors in their calculations.
- Recall or recognise most of the standard models that are needed and usually select appropriate ones to represent a variety of situations in the real world.

- Often correctly refer results from calculations using the model to the original situation; they sometimes give sensible interpretations of their results in context of the original realistic situation.
- Sometimes make intelligent comments on the modelling assumptions and possible refinements to the model.
- Comprehend or understand the meaning of most translations into mathematics of common realistic contexts.
- Often correctly refer the results of calculations back to the given context and sometimes make sensible comments or predictions.
- Distil much of the essential mathematical information from extended pieces of prose having mathematical content.
- Give some useful comments on this mathematical information.
- Usually make appropriate and effective use of contemporary calculator technology and other permitted resources, and are sometimes aware of any limitations to their use.
- Usually present results to an appropriate degree of accuracy.

Grade E candidates

- Recall or recognise some of the mathematical facts, concepts and techniques that are needed, and sometimes select appropriate ones to use in some contexts.
- Manipulate mathematical expressions and use graphs, sketches and diagrams, all with some accuracy and skill.
- Sometimes use mathematical language correctly and occasionally proceed logically through extended arguments or proofs.
- Recall or recognise some of the standard models that are needed and sometimes select appropriate ones to represent a variety of situations in the real world.
- Sometimes correctly refer results from calculations using the model to the original situation; they try to interpret their results in the context of the original realistic situation.
- Sometimes comprehend or understand the meaning of translations into mathematics of common realistic contexts.
- Sometimes correctly refer the results of calculations back to the given context and attempt to give comments or predictions.
- Distil some of the essential mathematical information from extended pieces of prose having mathematical content; they attempt to comment on this mathematical information.
- Candidates often make appropriate and efficient use of contemporary calculator technology and other permitted resources.
- Often present results to an appropriate degree of accuracy.

The table below shows how your uniform standardised mark is translated.

uniform mark (max. 100)	*80*	*70*	*60*	*50*	*40*
grade	*A*	*B*	*C*	*D*	*E*

The A* will be awarded to students who have achieved a grade A overall (480 UMS or more) and 180 UMS or more on the total of their Core 3 and Core 4 units. It is awarded for A Level qualification only and not for the AS qualification or individual units.

Four steps to successful revision

Step 1: Understand

- Study the topic to be learned slowly. Make sure you understand the logic or important concepts.
- Mark up the text if necessary – underline, highlight and make notes.
- Re-read each paragraph slowly.

GO TO STEP 2

Step 2: Summarise

- Now make your own revision note summary:
  What is the main idea, theme or concept to be learned?
  What are the main points? How does the logic develop?
  Ask questions: Why? How? What next?
- Use bullet points, mind maps, patterned notes.
- Link ideas with mnemonics, mind maps, crazy stories.
- Note the title and date of the revision notes
  (e.g. Mathematics: Trigonometry, 3rd March).
- Organise your notes carefully and keep them in a file.

This is now in **short-term memory**. You will forget 80% of it if you do not go to Step 3.
GO TO STEP 3, but first take a 10 minute break.

Step 3: Memorise

- Take 25 minute learning 'bites' with 5 minute breaks.
- After each 5 minute break test yourself:
  Cover the original revision note summary
  Write down the main points
  Speak out loud (record yourself)
  Tell someone else
  Repeat many times.

The material is well on its way to **long-term memory**.
You will forget 40% if you do not do step 4. *GO TO STEP 4*

Step 4: Track/Review

- Create a Revision Diary (one A4 page per day).
- Make a revision plan for the topic, e.g. 1 day later, 1 week later, 1 month later.
- Record your revision in your Revision Diary, e.g.
  Mathematics: Trigonometry, 3rd March 25 minutes
  Mathematics: Trigonometry, 5th March 15 minutes
  Mathematics: Trigonometry, 3rd April 15 minutes
  ... and then at monthly intervals.

Chapter 1

Core 1 Pure Mathematics

The following topics are covered in this chapter:

- *Algebra and functions*
- *Coordinate geometry in the (x, y) plane*
- *Sequences and series*
- *Differentiation*
- *Integration*

Note: Calculators are *not* allowed in this module

1.1 Algebra and functions

LEARNING SUMMARY

After studying this section you should be able to:

- *work with indices and surds*
- *use function notation*
- *solve quadratic equations and sketch the graphs of quadratic functions*
- *understand the definition of a polynomial*
- *recognise and sketch graphs of a range of functions*
- *use transformations to sketch graphs of related functions*
- *solve simultaneous equations – one linear and one quadratic*
- *solve linear and quadratic inequalities*

Indices

You need to know these basic rules and be able to apply them.

$a^m \times a^n = a^{m+n}$	$a^m \div a^n = a^{m-n}$	$(a^m)^n = a^{mn}$	$a^{\frac{1}{n}} = \sqrt[n]{a}$
$a^{\frac{m}{n}} = (a^m)^{\frac{1}{n}} = (a^{\frac{1}{n}})^m$	$a^{-n} = \frac{1}{a^n}$	$(ab)^n = a^n b^n$	$\left(\frac{a}{b}\right)^n = \frac{a^n}{b^n}$

Remember that, in general $(a+b)^n \neq a^n + b^n$ and $(a-b)^n \neq a^n - b^n$.

Some important special cases are: $a^1 = a$ $a^{\frac{1}{2}} = \sqrt{a}$

and $a^0 = 1$ $a^{-1} = \frac{1}{a}$ provided $a \neq 0$.

For example $9^{\frac{1}{2}} = \sqrt{9} = 3$ and $16^{\frac{3}{2}} = (16^{\frac{1}{2}})^3 = 4^3 = 64$.

Surds

A **surd** is the root of a whole number that has an **irrational** value.

Some examples are $\sqrt{2}$, $\sqrt{3}$ and $\sqrt{10}$.

You can often simplify a surd using the fact that $\sqrt{ab} = \sqrt{a} \times \sqrt{b}$ and choosing a or b to be a square number.

For example $\sqrt{12} = \sqrt{4 \times 3} = \sqrt{4} \times \sqrt{3} = 2\sqrt{3}$ and $\sqrt{18} + 2\sqrt{32} = 3\sqrt{2} + 2 \times 4\sqrt{2} = 11\sqrt{2}$.

An irrational number continues for ever after the decimal point without a repeating pattern.

To simplify $\frac{3}{\sqrt{5}}$, multiply the numerator and the denominator by $\sqrt{5}$ to get

$\frac{3}{\sqrt{5}} \times \frac{\sqrt{5}}{\sqrt{5}} = \frac{3\sqrt{5}}{5}$. This is called **rationalising the denominator**.

Functions

A **function** may be thought of as a rule which takes each member x of a set and assigns, or **maps**, it to some value y known as its **image**.

x maps to y
y is the image of x.

$$x \longrightarrow \boxed{\text{Function}} \longrightarrow y$$

$f(x)$ is read as 'f of x'.

A letter such as f, g or h is often used to stand for a function. The function which squares a number and adds on 5, for example, can be written as $f(x) = x^2 + 5$. The same notation may also be used to show how a function affects particular values. For this function, $f(4) = 4^2 + 5 = 21$, $f(-10) = (-10)^2 + 5 = 105$ and so on.

An alternative notation for the same function is $f\colon x \mapsto x^2 + 5$.

Quadratics

In algebra, any expression of the form $ax^2 + bx + c$ where $a \neq 0$ is called a **quadratic expression.**

You need to be able to expand brackets, **for example**

$$(2x + 3)(x - 4) = 2x^2 - 8x + 3x - 12$$
$$= 2x^2 - 5x - 12$$

It is useful to remember these special results:

$$(x + a)^2 = x^2 + 2ax + a^2 \qquad (x - a)^2 = x^2 - 2ax + a^2 \qquad (x + a)(x - a) = x^2 - a^2$$

Some examples are:

$$(x + 3)^2 = x^2 + 6x + 9 \qquad (x - 5)^2 = x^2 - 10x + 25$$
$$(x + 7)(x - 7) = x^2 - 49 \qquad (x + \sqrt{3})(x - \sqrt{3}) = x^2 - 3$$

To solve problems in algebra you need to develop your skills so that you can recognise how to apply the basic results.

In this fraction, the denominator $2 - \sqrt{3}$ is irrational.

This is an example of rationalising the denominator.

For example, the fraction $\dfrac{1}{2 - \sqrt{3}}$ may be simplified by multiplying the numerator and denominator by $2 + \sqrt{3}$ to make an equivalent fraction in which the denominator is rational.

$$\frac{1}{2 - \sqrt{3}} = \frac{1}{2 - \sqrt{3}} \times \frac{2 + \sqrt{3}}{2 + \sqrt{3}}$$
$$= \frac{2 + \sqrt{3}}{4 - 3} \qquad \text{(The denominator is now rational)}$$
$$= 2 + \sqrt{3}$$

$(2 - \sqrt{3})(2 + \sqrt{3})$
$= 2^2 - (\sqrt{3})^2$
$= 4 - 3.$

Quadratic equations

Equations of the form $ax^2 + bx + c = 0$ (where $a \neq 0$) are **quadratic equations.**

Some quadratic equations can be solved by **factorising** the quadratic expression.

Factorise.

Example Solve $2x^2 - x - 3 = 0$.

$(2x - 3)(x + 1) = 0$

Either $2x - 3 = 0$ or $x + 1 = 0$

so $x = \frac{3}{2}$ or $x = -1$.

Some equations can be converted into a quadratic equation by substitution.

Example Solve $x^{\frac{2}{3}} + x^{\frac{1}{3}} - 6 = 0$

Substituting $y = x^{\frac{1}{3}}$ gives

$y^2 + y - 6 = 0$

$(y + 3)(y - 2) = 0$

Either $y + 3 = 0$ or $y - 2 = 0$

so $y = -3$ or $y = 2$.

Re-writing in terms of x gives $x^{\frac{1}{3}} = -3$ or $x^{\frac{1}{3}} = 2$,

so $x = -27$ or $x = 8$.

If the quadratic will not factorise then you can try **completing the square**.

$x^2 - 6x = (x - 3)^2 - 9$ so $x^2 - 6x + 1 = (x - 3)^2 - 8$.

$\sqrt{8} = \sqrt{(4 \times 2)} = 2\sqrt{2}$.

Example Solve $x^2 - 6x + 1 = 0$

$(x - 3)^2 - 8 = 0$ (Now x only appears once in the equation)

$(x - 3)^2 = 8$

$x - 3 = \pm\sqrt{8}$

$x = 3 \pm 2\sqrt{2}$.

(The solutions are $3 + 2\sqrt{2}$ and $3 - 2\sqrt{2}$)

The method shown for completing the square can be adapted for the general form of a quadratic. This gives the **quadratic formula**.

$ax^2 + bx + c = 0$

Multiply both sides by $4a$.

$4a^2x^2 + 4abx + 4ac = 0$ Note: $4a^2x^2 + 4abx + b^2 = (2ax + b)^2$.

Add b^2 to complete the square and subtract it again to keep things the same.

$4a^2x^2 + 4abx + b^2 + 4ac - b^2 = 0$

$(2ax + b)^2 + 4ac - b^2 = 0$ Now x only appears once in the equation.

$(2ax + b)^2 = b^2 - 4ac$

Find the square root of both sides.

$2ax + b = \pm\sqrt{b^2 - 4ac}$

Rearrange to find x.

$$x = \frac{-b \pm \sqrt{b^2 - 4ac}}{2a}$$

You must learn this formula.

Example Solve $3x^2 - 4x - 2 = 0$ giving your answers in surd form.

Comparing this equation with the general form gives $a = 3$, $b = -4$ and $c = -2$. Substitute this information into the formula:

In Core 1 you do not have access to a calculator so you will have to leave the answer in surd form.

$$x = \frac{4 \pm \sqrt{(-4)^2 - 4(3)(-2)}}{6} = \frac{4 \pm \sqrt{40}}{6} = \frac{4 \pm 2\sqrt{10}}{6} = \frac{2 \pm \sqrt{10}}{3}$$

so $x = \dfrac{2 + \sqrt{10}}{3}$ or $x = \dfrac{2 - \sqrt{10}}{3}$.

Discriminant

In the formula, the value of $b^2 - 4ac$ is called the **discriminant**. This value can be used to give information about the solutions without having to solve the equation.

In the previous example, $b^2 - 4ac = 40 > 0$ and two real roots were found.

$b^2 - 4ac > 0$ *two* distinct real solutions (the solutions are often called **roots**)

$b^2 - 4ac = 0$ *one* real solution (often thought of as a **repeated root**)

$b^2 - 4ac < 0$ *no* real solutions.

The solutions of a quadratic equation correspond to where the graph of the quadratic function crosses the x-axis. There are three possible situations depending on the value of the discriminant.

The diagrams correspond to the situation where $a > 0$. The same principle applies when $a < 0$ but the diagrams appear the other way up.

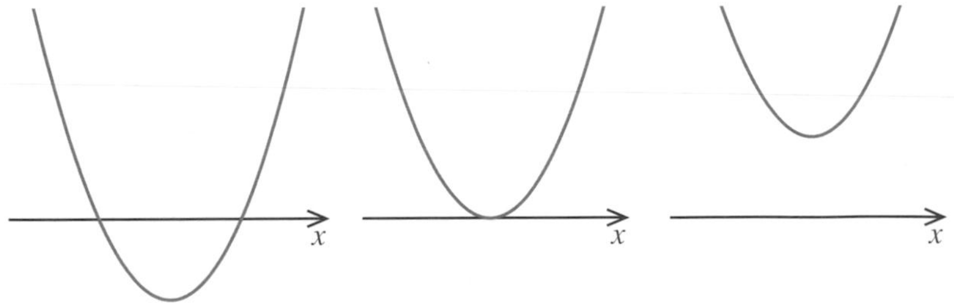

$b^2 - 4ac > 0$ $b^2 - 4ac = 0$ $b^2 - 4ac < 0$

Example The equation $5x^2 + 3x + p = 0$ has a repeated root. Find the value of p.

In this case, $a = 5$, $b = 3$ and $c = p$.

For a repeated root $b^2 - 4ac = 0$ so $9 - 20p = 0$, giving $p = 0.45$

Quadratic graphs

The graph of $y = ax^2 + bx + c$ is a parabola.

$a > 0$ $a < 0$

The methods used for solving quadratic equations can also be used to give information about the graphs.

Example Sketch the graph of $y = x^2 - x - 6$.
Find the coordinates of the lowest point on the curve.

The curve will cross the x-axis when $y = 0$. You can find these points by solving the equation $x^2 - x - 6 = 0$.

The curve will cross the y-axis when $x = 0$, giving $y = -6$.

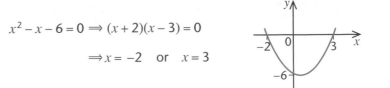

$$x^2 - x - 6 = 0 \Rightarrow (x + 2)(x - 3) = 0$$
$$\Rightarrow x = -2 \quad \text{or} \quad x = 3$$

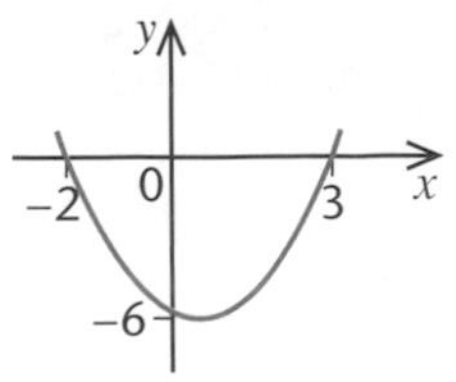

The curve is symmetrical so the lowest point occurs mid-way between –2 and 3 and this is given by $(-2 + 3) \div 2 = 0.5$

When $x = 0.5$, $y = 0.5^2 - 0.5 - 6 = -6.25$

The vertex of the curve occurs at its maximum or minimum point.

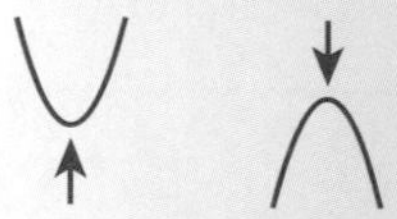

The lowest point on the curve is (0.5, –6.25).

Completing the square gives information about the **vertex** of the curve even if the equation will not factorise.

Example Find the coordinates of the vertex of the curve $y = x^2 + 2x + 3$.

You need to recognise that $x^2 + 2x + 1 = (x + 1)^2$,

then, completing the square, $x^2 + 2x + 3 = x^2 + 2x + 1 + 2 = (x + 1)^2 + 2$.

The equation of the curve can now be written as $y = (x + 1)^2 + 2$.

$(x + 1)^2$ cannot be negative so its minimum value is zero, when $x = -1$.

This means that the minimum value of y is 2 and this occurs when $x = -1$.

The vertex of the curve is at (–1, 2).

Polynomials

A quadratic is one example of a **polynomial**. In general, a polynomial takes the form:

$$a_n x^n + a_{n-1} x^{n-1} + a_{n-2} x^{n-2} + \ldots + a_0,$$

This is much simpler than it looks.

where $a_n, a_{n-1}, \ldots a_0$ are constants and n is a positive whole number.

For example, $x^4 - 2x^3$ is a polynomial. In this case $a_4 = 1$, $a_3 = -2$ and a_2, a_1, a_0 are all zero.

The **degree** of a polynomial is the highest power of x that it includes, so the degree of $x^4 - 2x^3$ is 4. A quadratic is a polynomial of degree 2, a cubic is a polynomial of degree 3 and so on.

Higher order polynomials can also be factorised. **For example**, $x^3 + 6x^2 + 11x + 6 = (x + 1)(x + 2)(x + 3)$

More graphs

The graph of a cubic function $y = ax^3 + bx^2 + cx + d$ can take a number of forms.

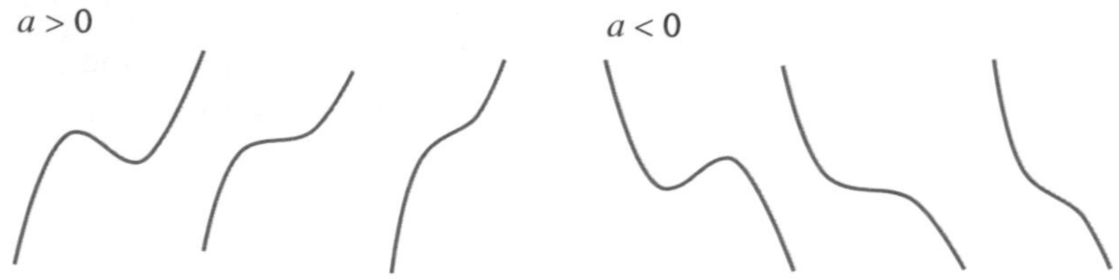

KEY POINT

The graph of a cubic function that can be factorised as $y = (x - p)(x - q)(x - r)$ will cross the x-axis at p, q and r. If any two of p, q and r are the same then the x-axis will be a tangent to the curve at that point.

For example, the graph of $y = (x+2)(x-3)^2$ looks like this:

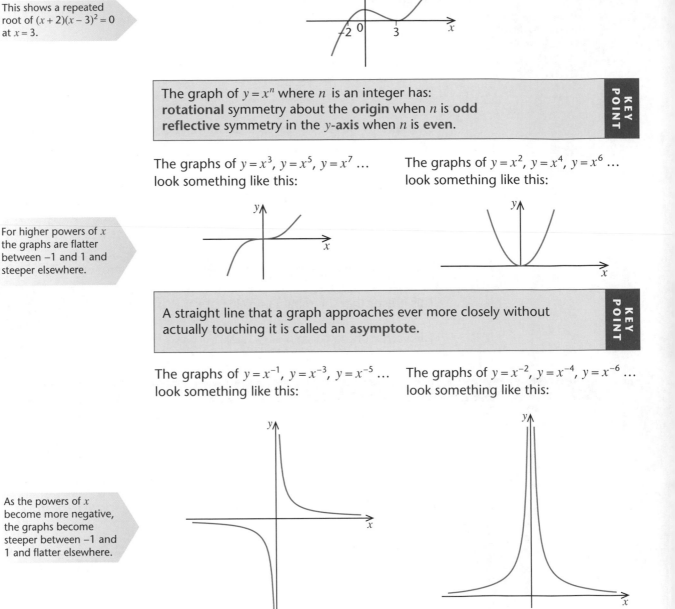

This shows a repeated root of $(x+2)(x-3)^2 = 0$ at $x = 3$.

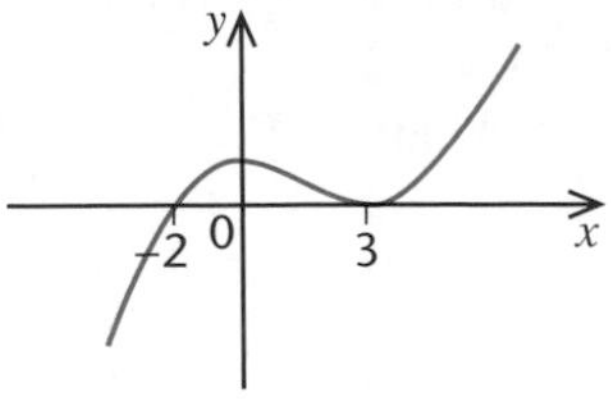

> **KEY POINT**
> The graph of $y = x^n$ where n is an integer has:
> **rotational** symmetry about the **origin** when n is **odd**
> **reflective** symmetry in the **y-axis** when n is **even**.

The graphs of $y = x^3$, $y = x^5$, $y = x^7$ … look something like this:

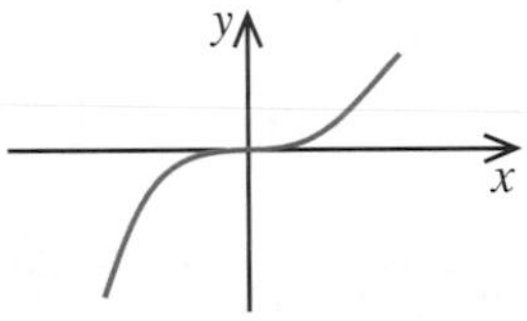

The graphs of $y = x^2$, $y = x^4$, $y = x^6$ … look something like this:

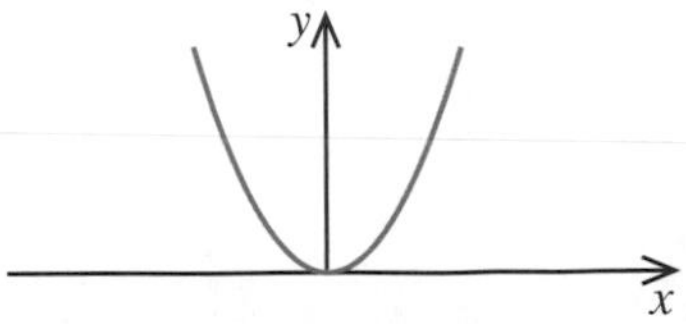

For higher powers of x the graphs are flatter between –1 and 1 and steeper elsewhere.

> **KEY POINT**
> A straight line that a graph approaches ever more closely without actually touching it is called an **asymptote**.

The graphs of $y = x^{-1}$, $y = x^{-3}$, $y = x^{-5}$ … look something like this:

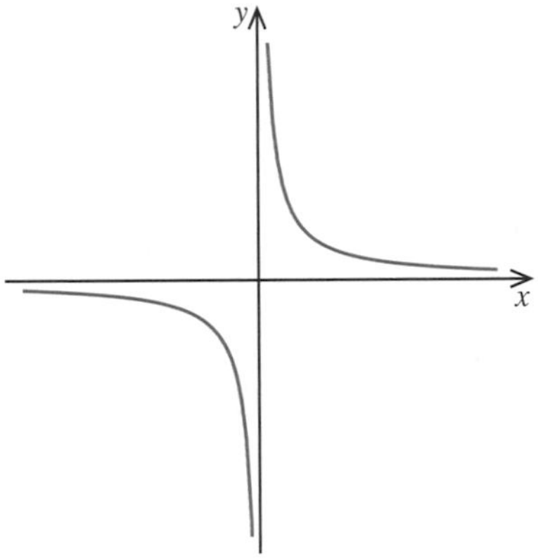

Both the x and y-axes are asymptotes for these graphs.

The graphs of $y = x^{-2}$, $y = x^{-4}$, $y = x^{-6}$ … look something like this:

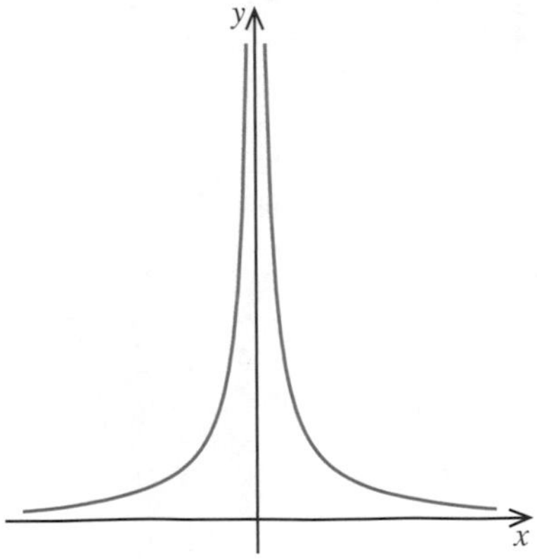

The x-axis and the positive y-axis are asymptotes for these graphs.

As the powers of x become more negative, the graphs become steeper between –1 and 1 and flatter elsewhere.

The graphs of $y = k\sqrt{x}$, $k > 0$, look something like this. These graphs do not have any asymptotes.

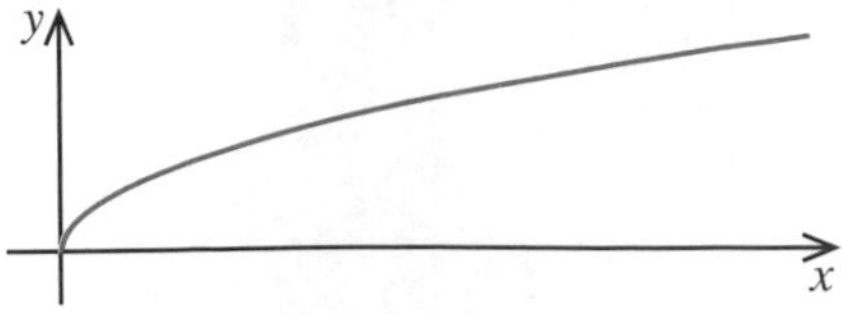

Transforming graphs

The graph of some new function can often be obtained from the graph of a known function by applying a transformation. A summary of the standard transformations is given in the table.

Known function	*New function*	*Transformation*
$y = f(x)$	$y = f(x) + a$	Translation through a units parallel to y-axis.
	$y = f(x - a)$	Translation through a units parallel to x-axis.
	$y = af(x)$	One-way stretch with scale factor a parallel to the y-axis.
	$y = f(ax)$	One-way stretch with scale factor $\frac{1}{a}$ parallel to the x-axis.

You may need to apply a combination of transformations in some cases.

Example The diagram shows the graph of a function, $y = f(x)$ for $1 \leqslant x \leqslant 3$

Use the same axes to show:

(a) $y = f(x) + 1$
(b) $y = f(x + 1)$
(c) $y = 2f(x)$
(d) $y = f(2x)$

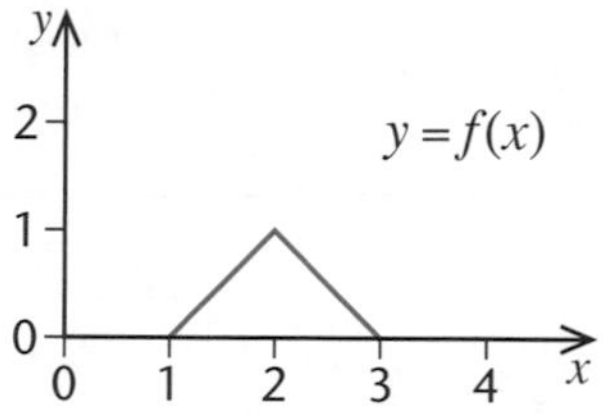

(a)

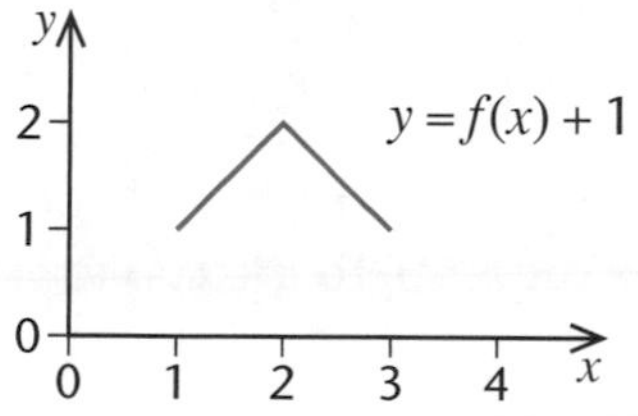

(b)

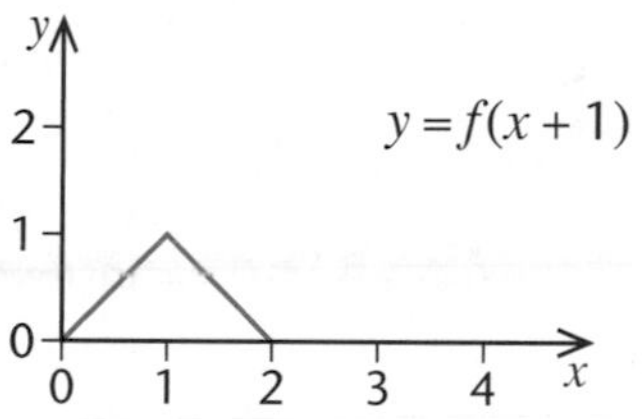

(c)

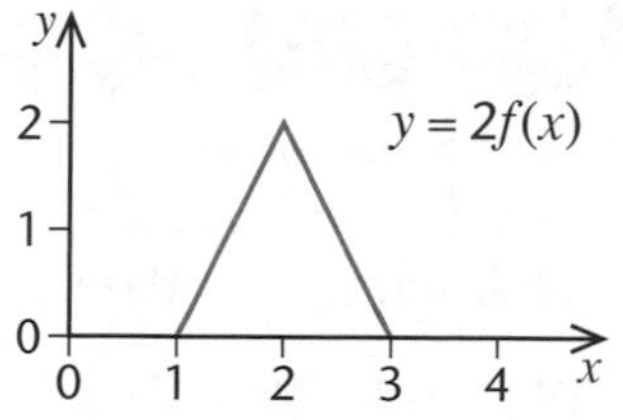

(d)

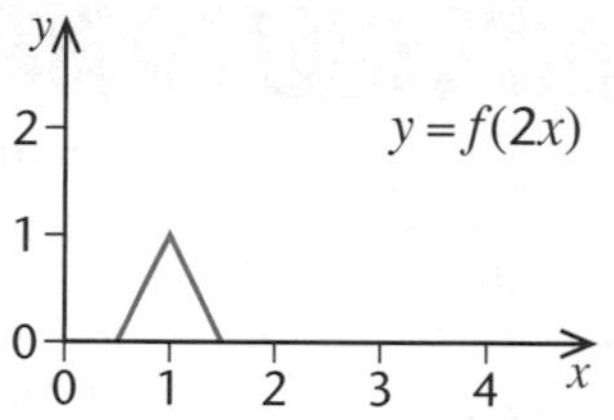

Simultaneous equations

The substitution method of solving pairs of linear simultaneous equations can also be applied when one of the equations is a quadratic.

Example Find the coordinates of the points where the line $y = x + 1$ intersects the circle $x^2 + y^2 = 5$.

$$y = x + 1 \quad (1)$$
$$x^2 + y^2 = 5 \quad (2)$$

Substitute for y from (1) into (2): $x^2 + (x + 1)^2 = 5$.

Now expand the brackets: $x^2 + x^2 + 2x + 1 = 5$.

Arrange in the form $ax^2 + bx + c = 0$: $2x^2 + 2x - 4 = 0$.

Divide by 2.

$x^2 + x - 2 = 0$

Solve the equation: $(x - 1)(x + 2) = 0$

Either $x = 1$ or $x = -2$

Substitute into (1) to find the y values: When $x = 1$, $y = 2$

When $x = -2$, $y = -1$

The coordinates of the points of intersection are (1, 2) and (−2, −1).

Geometrical interpretation of algebraic solutions

In the above example, when $y = x + 1$ and $x^2 + y^2 = 5$ are solved simultaneously, the resulting quadratic equation in x has two distinct solutions. This gives the two points of intersection of the line and the curve.

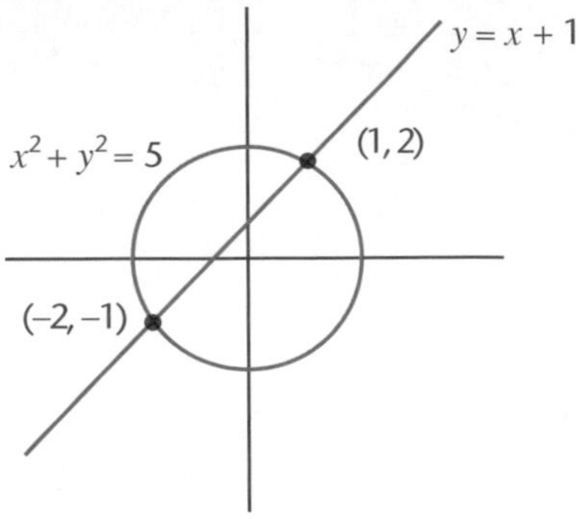

When you solve the equations of a line and a curve simultaneously and form a quadratic equation, $ax^2 + bx + c = 0$, the discriminant, $b^2 - 4ac$, gives information about the number of points of intersection.

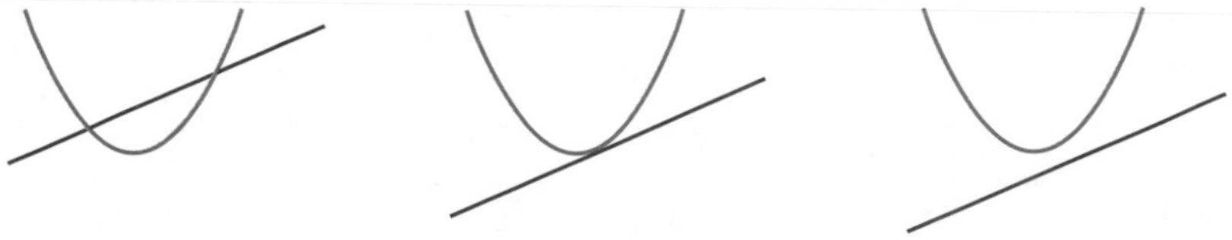

If $b^2 - 4ac > 0$ the line and curve intersect in two distinct points.

If $b^2 - 4ac = 0$ the line is a tangent to the curve.

If $b^2 - 4ac < 0$ the line and the curve do not intersect.

Inequalities

Linear inequalities can be solved by rearrangement in much the same way as linear equations. However, care must be taken to reverse the direction of the inequality when multiplying or dividing by a negative.

Example Solve the inequality $8 - 3x > 23$

Subtract 8 from both sides: $-3x > 15$
Divide both sides by -3: $x < -5$

An inequality which has x on both sides is treated like the corresponding equation.

Example Solve the inequality $5x - 3 > 3x - 10$

Subtract $3x$ from both sides: $2x - 3 > -10$
Add 3 to both sides: $2x > -7$
Divide both sides by 2: $x > -3.5$

Quadratic inequalities are solved in a similar way to quadratic equations but a sketch graph is often helpful at the final stage.

Example Solve the inequality $x^2 - 3x + 2 < 0$.

Factorise the quadratic expression: $(x - 1)(x - 2) < 0$.

$y = 0$ when $x = 1$ or when $x = 2$. The graph must cross the x-axis at these points.

Sketch the graph of $y = (x - 1)(x - 2)$:

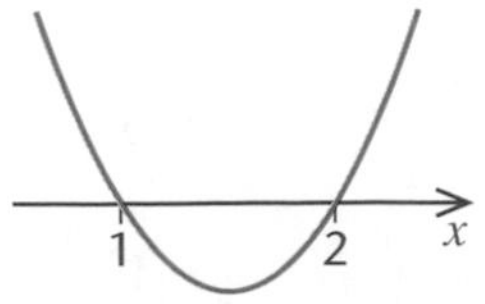

$(x - 1)(x - 2) < 0$ when the curve is below the x-axis.

The graph shows that $(x - 1)(x - 2) < 0$ for x values between 1 and 2.

It follows that $x^2 - 3x + 2 < 0$ when $1 < x < 2$.

If the quadratic expression cannot be factorised then the formula may be used to find the points of intersection of the curve with the x-axis.

Example Solve the inequality $x^2 + 2x - 5 > 0$.

$\sqrt{24} = \sqrt{4 \times 6} = 2\sqrt{6}$

Use the formula to solve $x^2 + 2x - 5 = 0$. $x = \dfrac{-2 \pm \sqrt{24}}{2} = \dfrac{-2 \pm 2\sqrt{6}}{2}$

Simplify the result. $x = -1 \pm \sqrt{6}$

$x^2 + 2x - 5 > 0$ when the curve is above the x-axis.

Sketch the graph of $y = x^2 + 2x - 5$:

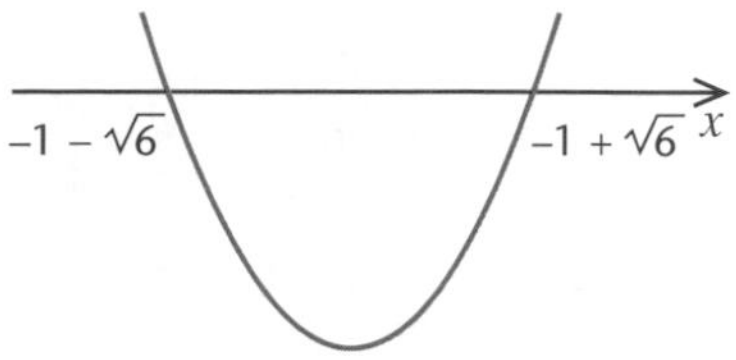

Write the solution as two separate inequalities.

From the sketch, $x^2 + 2x - 5 > 0$ when $x < -1 - \sqrt{6}$ or when $x > -1 + \sqrt{6}$.

Progress check

1 Simplify these surd expressions:
(a) $\sqrt{72}$ (b) $(1 + \sqrt{3})(1 - \sqrt{3})$ (c) $5\sqrt{12} - 6\sqrt{3}$.

2 (a) Solve $2x^2 + x - 21 = 0$ by factorising.
(b) Solve $x^2 - 6x + 7 = 0$ by completing the square. Give the roots in surd form.
(c) Solve $3x^2 + 2x - 2 = 0$ using the quadratic formula. Give the roots in surd form.

3 Express $x^2 - 8x + 17$ in the form $(x - a)^2 + b$ and hence write down its minimum value.

4 Sketch the graph of $y = (x + 3)(x - 2)^2$.

5 The diagram shows the graph of $y = f(x)$ for $0 \leqslant x \leqslant 2$.
Sketch the graph of $y = f(x - 2)$

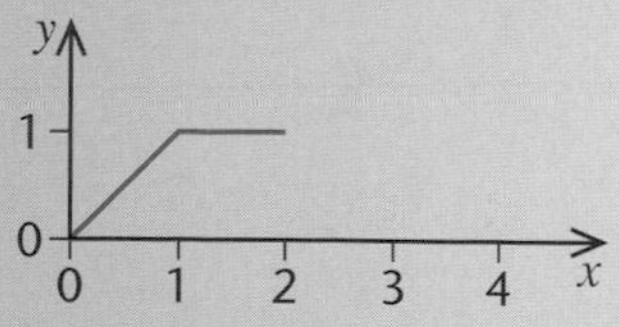

6 Solve these equations simultaneously.
$x + 2y = 6$
$4y^2 - 5x^2 = 36$.

7 Find the values of A, B and C.
$3x^3 - 16x - 1 \equiv Ax(x - 1)(x + 2) + B(x - 1)^2 + C(4x^2 + 3)$.

Progress check *(continued)*

8 Solve these inequalities.

(a) $3x - 7 > 5x + 11$ (b) $2x^2 - 3x - 5 \leqslant 0$.

1 (a) $6\sqrt{2}$ (b) -2 (c) $4\sqrt{3}$

2 (a) $3, -3.5$ (b) $3 \pm \sqrt{2}$ (c) $\dfrac{-1 \pm \sqrt{7}}{3}$

3 $(x - 4)^2 + 1$, 1

4

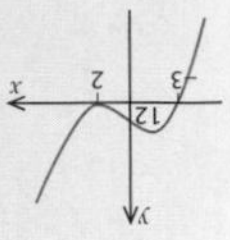

5

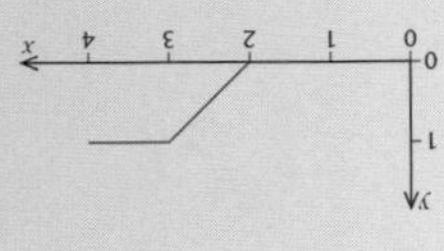

6 $x = 0, y = 3$ or $x = -3, y = 4.5$

7 $A = 3, B = 5, C = -2$

8 (a) $x < -9$ (b) $-1 \leqslant x \leqslant 2.5$

1.2 Coordinate geometry in the (x, y) plane

LEARNING SUMMARY

After studying this section you should be able to:

- *find the gradient of a line joining two points*
- *recognise equations of straight lines in various forms*
- *construct the equation of a straight line parallel to a given line and passing through a given point*
- *construct the equation of a straight line perpendicular to a given line and passing through a given point*
- *construct the equation of a straight line passing through two given points*
- *find the coordinates of the mid-point of two given points*

Gradient of a line

KEY POINT

The gradient of a line joining the points (x_1, y_1) and (x_2, y_2) is given by the formula $m = \dfrac{y_2 - y_1}{x_2 - x_1}$.

For example, the gradient of the line joining (2, –5) and (–1, 4) is $\dfrac{4-(-5)}{-1-2} = \dfrac{9}{-3} = -3$.

Straight lines

The general equation of a straight line is $ax + by + c = 0$. The equation of any straight line can be written in this form or as $y = mx + c$.

For example, the line $x + y = 5$ corresponds to $a = 1$, $b = 1$ and $c = -5$.

The gradient of a vertical line is undefined.

Straight lines, apart from those parallel to the y-axis, can be written in the form $y = mx + c$. This is known as **gradient–intercept** form because the gradient (m) and the y-intercept (c) are clearly identified in the equation. This makes it easy to construct an equation when the gradient and intercept are known.

For example, the line with gradient 4 crossing the y-axis at –5 has equation $y = 4x - 5$.

KEY POINT

Straight lines that are **parallel** must have the **same gradient**.

Example Find the equation of the straight line parallel to $y = 3x - 5$ and passing through the point (4, 2).

This equation must be satisfied at the point where $x = 4$ and $y = 2$.

The line must have gradient 3 and so it can be written in the form $y = 3x + c$.
Substituting $x = 4$ and $y = 2$ gives $2 = 3 \times 4 + c \Rightarrow c = -10$.
The required equation is $y = 3x - 10$.

This result is used frequently to find the equation of a tangent or a normal to a curve at a given point.

KEY POINT
The equation of a straight line with gradient m and passing through the point (x_1, y_1) can be written as $y - y_1 = m(x - x_1)$.

Using this in the example above gives $y - 2 = 3(x - 4)$. The equation is acceptable in this form but it can be rearranged to give $y = 3x - 10$ as before.

Example Find the equation of the straight line passing through the points (–1,5) and (3,–2).

One approach is to use the form $y = mx + c$ to produce a pair of simultaneous equations:

Substituting $x = -1$ and $y = 5$ gives $\quad 5 = -m + c \quad (1)$
Substituting $x = 3$ and $y = -2$ gives $\quad -2 = 3m + c \quad (2)$

(2) – (1) gives: $\quad -7 = 4m \Rightarrow m = \dfrac{-7}{4}$.

Substituting for m in (1) gives: $\quad 5 = \dfrac{7}{4} + c \Rightarrow c = \dfrac{13}{4}$.

The equation of the line is $y = \dfrac{-7}{4}x + \dfrac{13}{4}$. This is the same as $4y + 7x = 13$.

An alternative approach is to find the gradient directly and then use the form $y - y_1 = m(x - x_1)$.

Taking (–1, 5) as (x_1, y_1) and (3, –2) as (x_2, y_2),

$$m = \frac{-2-5}{3-(-1)} = \frac{-7}{4} = -\frac{7}{4}.$$

You can use either of the points (–1, 5) or (3, –2).

The equation of the line is $y - 5 = -\dfrac{7}{4}(x + 1)$.

To show that this is the same as the previous result (you don't *need* to do this):

Multiply both sides of the equation by 4: $\quad 4y - 20 = -7(x + 1)$.
Expand the brackets: $\quad 4y - 20 = -7x - 7$.
Rearrange to give the previous result: $\quad 4y + 7x = 13$.

An equation of the form $ax + by + c = 0$ can be rearranged into gradient–intercept form provided that $b \neq 0$. This becomes $y = -\frac{a}{b}x - \frac{c}{b}$ and shows that parallel lines may be produced by keeping a and b fixed and allowing c to change.

Note that $4x - 6y = 7$ is equivalent to $2x - 3y = 3.5$

For example, the lines $2x - 3y = 5$, $2x - 3y = -2$, $4x - 6y = 7$ are all parallel.

$m_1 \times m_2 = -1$.

KEY POINT
When two straight lines are **perpendicular**, the **product** of their **gradients is –1**.

Example Find the equation of the straight line perpendicular to the line $4x + 3y = 12$ and passing through the point (2, 5).

Rearrange the equation into gradient–intercept form: $\quad y = -\frac{4}{3}x + 4$.
If the gradient of the required line is m then: $\quad m \times -\frac{4}{3} = -1 \Rightarrow m = \frac{3}{4}$
Using the form $y - y_1 = m(x - x_1)$ gives: $\quad y - 5 = \frac{3}{4}(x - 2)$.

> **KEY POINT**
>
> If P has coordinates (x_1, y_1) and Q has coordinates (x_2, y_2) then the mid-point of PQ has coordinates $\left(\dfrac{x_1 + x_2}{2}, \dfrac{y_1 + y_2}{2}\right)$.

For example, the mid-point of the line joining (–3, 1) and (5, 7) is $\left(\dfrac{-3+5}{2}, \dfrac{1+7}{2}\right) = (1, 4)$.

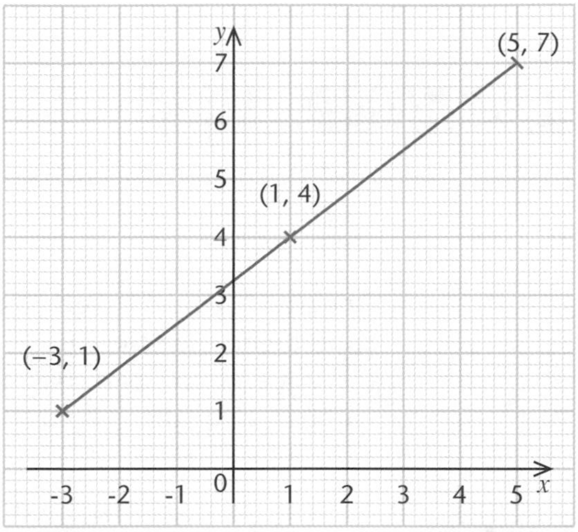

Progress check

> The parallel line can be written in the form $3x + 2y = C$. To find C just substitute the coordinates (5, 4) for x and y.

1 The equation of a straight line is $y = 2x - 7$.
 (a) Write down the equation of a parallel line crossing the y-axis at 5.
 (b) Find the equation of a parallel line passing through the point (4, 9).

2 Find the equation of the straight line joining the points (–2,–9) and (1, 3).

3 The equation of a straight line is $3x + 2y = 11$.
 (a) Find the equation of a parallel line passing through the point (5, 4).
 (b) Find the equation of a perpendicular line passing through the point (–2, 6).

4 Find the coordinates of the mid-point of (–9, 3) and (–5, 1).

5 P is the point (–1,–4) and Q is the point (5,–1). Find the equation of the line perpendicular to PQ and passing through its mid-point.

1 (a) $y = 2x + 5$ (b) $y = 2x + 1$
2 $y = 4x - 1$
3 (a) $3x + 2y = 23$ (b) $y - 6 = \frac{2}{3}(x + 2)$ or $3y - 2x - 22 = 0$.
4 (–7, 2)
5 $y = -2x + 1.5$

1.3 Sequences and series

After studying this section you should be able to:

- *understand the different types of sequence, including the notation and formulae used to describe them*
- *recognise arithmetic progressions and calculate the sum of their series*

LEARNING SUMMARY

Sequences and series

A list of numbers in a particular order, that follow some rule for finding later values, is called a **sequence**. Each number in a sequence is called a **term**, and terms are often denoted by $u_1, u_2, u_3, \ldots, u_n, \ldots$.

One way to define a sequence is to give a formula for the nth term such as $u_n = n^2$. Substituting the values $n = 1, 2, 3, 4, \ldots$ produces the sequence 1, 4, 9, 16, … and the value of any particular term can be calculated by substituting its position number into the formula. For example, in this case, the 50th term is $u_{50} = 50^2 = 2500$.

Another way to define a sequence is to give a starting value together with a rule that shows the connection between successive terms. This is sometimes called a **recursive definition**. For example $u_1 = 5$ and $u_{n+1} = 2u_n$ defines the sequence 5, 10, 20, 40, … . The rule $u_{n+1} = 2u_n$ is an example of a **recurrence relation**.

Two special sequences are the **arithmetic progression** (A.P.) and the **geometric progression** (G.P.). Geometric progression is studied in Core 2.

In an A.P. successive terms have a **common difference**, e.g. 1, 4, 7, 10, … . The first term is denoted by a and the common difference is d. With this notation, the definition of an A.P. may now be given as $u_1 = a,\ u_{n+1} = u_n + d$.

The terms of an A.P. take the form $a,\ a + d,\ a + 2d,\ a + 3d, \ldots$ and the nth term is given by $u_n = a + (n - 1)d$..

A **series** is formed by adding together the terms of a sequence. The use of sigma notation can greatly simplify the way that series are written. For example, the series $1^2 + 2^2 + 3^2 + \ldots + n^2$ may be written as $\sum_{i=1}^{n} i^2$. The sum of the first n terms of a series is often denoted by S_n and so $S_n = u_1 + u_2 + u_3 + \ldots + u_n = \sum_{i=1}^{n} u_i$.

Arithmetic series

The **sum of an arithmetic series** is given by

$$S_n = a + (a + d) + (a + 2d) + \ldots + (a + (n - 1)d).$$

This may also be written as

$$S_n = l + (l - d) + (l - 2d) + \ldots + (l - (n - 1)d),$$

where l is the last term.

Adding the two versions gives

$$2S_n = (a + l) + (a + l) + (a + l) + \ldots (a + l)$$
$$= n(a + l).$$
$$S_n = \frac{n}{2}(a + l).$$

You need to know how to establish the general result.

KEY POINT

The sum of the first n terms of an arithmetic series is: $S_n = \frac{n}{2}(a + l)$.

Substituting $l = a + (n - 1)d$ gives the alternative form of the result:

$$S_n = \frac{n}{2}(2a + (n - 1)d).$$

A special case is the sum of the first n natural numbers:

$$1 + 2 + 3 + \ldots + n = \frac{n}{2}(n + 1).$$

For example, the sum of the first 100 natural numbers is $\frac{100}{2}(100 + 1) = 5050$.

When finding the sum of an A.P. you need to select the most appropriate version of the formula to suit the information.

Example Find the sum of the first 50 terms of the series 15 + 18 + 21 + 24 + ...

In this series, $a = 15$, $d = 3$ and $n = 50$.

Using $S_n = \frac{n}{2}(2a + (n - 1)d)$ gives $S_{50} = \frac{50}{2}(2 \times 15 + 49 \times 3) = 4425$.

Progress check

1 Write down the first five terms of the sequence given by:
(a) $u_n = 2n^2$ (b) $u_1 = 10$, $u_{n+1} = 3u_n + 2$.

2 Find the 20th term of an A.P. with first term 7 and common difference 5.

3 Find the sum of the first 1000 natural numbers.

1 (a) 2, 8, 18, 32, 50 (b) 10, 32, 98, 296, 890
2 102
3 500 500

1.4 Differentiation

After studying this section you should be able to:

LEARNING SUMMARY

- *differentiate functions of the form x^n*
- *use differentiation to find the exact value of the gradient of a curve*
- *find the equation of the tangent and normal to a curve at a given point*

Gradient function

The **gradient** of a curve changes continuously along its length. Its value at any point P is given by the gradient of the **tangent** to the curve at P. The gradient may be found approximately by drawing.

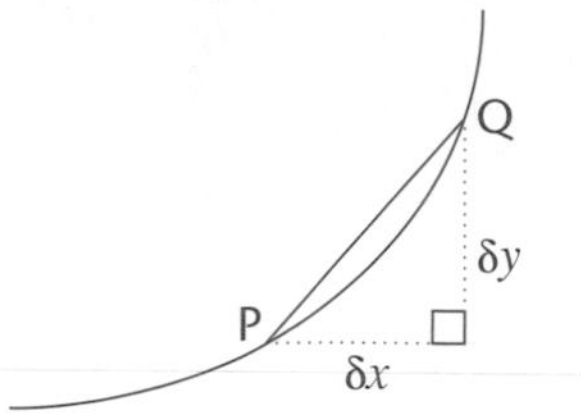

The exact value of the gradient is found by **differentiation**. This is the **limit** of the gradient of PQ as Q moves towards P.

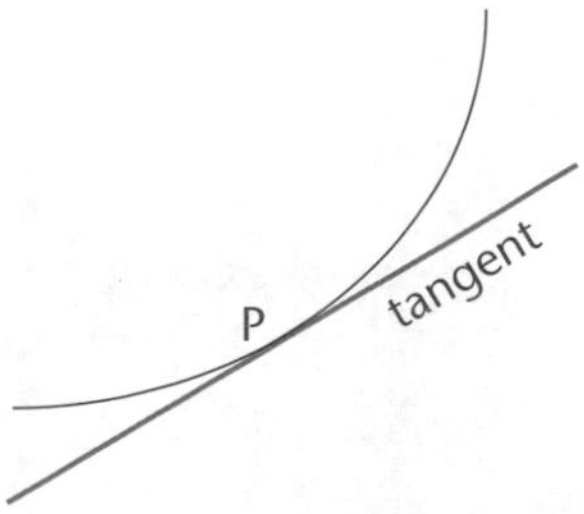

Gradient of curve at P $\approx \frac{\delta y}{\delta x}$.

As Q moves towards P, $\delta x \longrightarrow 0$ and $\frac{\delta y}{\delta x} \longrightarrow \frac{dy}{dx}$.

KEY POINT

> $\frac{dy}{dx}$ is the **gradient function** and represents the **derivative** of y with respect to x.
>
> The graph of $y = kx^n$ has gradient function $\frac{dy}{dx} = nkx^{n-1}$.

Example Find the gradient of the curve $y = 3x^2$ at the point P (5, 75).

$\frac{dy}{dx} = 2 \times 3x^1 = 6x$. At P, $x = 5$ so the gradient is $6 \times 5 = 30$.

Function notation may also be used for derivatives. If $y = f(x)$ then $\frac{dy}{dx} = f'(x)$.

The notation is useful for stating some of the rules of differentiation.

KEY POINT

> If $y = f(x) \pm g(x)$ then $\frac{dy}{dx} = f'(x) \pm g'(x)$

Example $f(x) = x^3 + 5x + 2$. Find (a) $f'(x)$ (b) $f'(4)$.

(a) $f'(x) = 3x^2 + 5$ (b) $f'(4) = 3 \times 4^2 + 5 = 53$.

Differentiate each term separately. Remember that $x^0 = 1$. When you differentiate a constant, the result is zero.

You often need to express a function in the right form before you can differentiate it.

Example Differentiate: (a) $f(x) = \sqrt{x}$ (b) $f(x) = (x - 5)(x + 3)$ (c) $f(x) = \frac{x^3 + 1}{x^2}$.

(a) $\sqrt{x}$ has to be written as a power of x to use the rule $f'(x) = nkx^{n-1}$.

$\sqrt{x} = x^{\frac{1}{2}}$ so $f(x) = x^{\frac{1}{2}}$ and $f'(x) = \frac{1}{2}x^{-\frac{1}{2}}$.

You cannot differentiate a product term by term.

(b) The brackets must first be removed and *then* you can differentiate term by term.

$f(x) = (x - 5)(x + 3) = x^2 - 2x - 15$.

$f'(x) = 2x - 2$.

You cannot differentiate a quotient term by term.

(c) Divide $x^3 + 1$ by x^2 first giving $f(x) = x + x^{-2}$.

Now, $f'(x) = 1 - 2x^{-3} = 1 - \dfrac{2}{x^3}$.

You also have to differentiate expressions when n is not a positive integer. You may have to change the expression into index form first.

Example 1 Differentiate $y = x^{\frac{3}{2}} + \dfrac{3}{x^2}$

Re-write as $y = x^{\frac{3}{2}} + 3x^{-2}$ and differentiate term by term to give

$$\frac{dy}{dx} = \frac{3}{2}x^{\frac{1}{2}} + (-2) \times 3x^{-3}$$
$$= \frac{3}{2}\sqrt{x} - \frac{6}{x^3}$$

Example 2 Differentiate $y = x\sqrt{x}$

Re-write as $y = x^{\frac{3}{2}}$ giving

$$\frac{dy}{dx} = \frac{3}{2}\sqrt{x}$$

Tangents and normals

You can find the gradient of the tangent to a curve at a point by differentiation. Then you can use the techniques described in the Coordinate Geometry section to find the **equation of the tangent** and the **equation of the normal** at the given point.

Example Find the equation of the tangent and the normal to the curve $y = x^3 - 4x$ at the point (2, 0).

Differentiate the equation of the curve to give $\dfrac{dy}{dx} = 3x^2 - 4$.

The gradient of the tangent at (2, 0) is $3 \times 2^2 - 4 = 8$.

Using $y - y_1 = m(x - x_1)$ gives the equation of the tangent as $y = 8(x - 2)$.

The normal at a point is perpendicular to the tangent.

The gradient of the normal is $-\frac{1}{8}$. Using $y - y_1 = m(x - x_1)$ again gives the equation of the normal as $y = -\frac{1}{8}(x - 2)$. You could rearrange this to give $x + 8y = 2$.

Progress check

1 Differentiate with respect to x:

$y = 6\sqrt{x}$.

2 Differentiate with respect to x:

$f(x) = \dfrac{2x + 1}{x^3}$.

3 Find the gradient of the curve $y = (4x - 1)^2$ at the point (1, 9).

4 Find the equation of: (a) the tangent (b) the normal to the curve $y = x^3 - 2x$ at the point (2, 4).

5 Find $\dfrac{dy}{dx}$ in each of the following questions.

(a) $y = x^{\frac{5}{3}} - \dfrac{1}{x}$

(b) $y = x^3\sqrt{x}$

1 $\dfrac{3}{\sqrt{x}}$

2 $-\dfrac{4}{x^3} - \dfrac{3}{x^4}$

3 24

4 (a) Tangent: $y = 10x - 16$; (b) normal: $x + 10y - 42 = 0$.

5 (a) $\frac{5}{3}x^{\frac{2}{3}} + \dfrac{1}{x^2}$

(b) $\frac{7}{2}x^{\frac{5}{2}}$

1.5 Integration

LEARNING SUMMARY

After studying this section you should be able to:

- *find an indefinite integral and understand what it represents*

Indefinite integration

The idea of a **reverse process** is an important one in many areas of mathematics. In **calculus**, the reverse process of differentiation is **integration** and this turns out to be an extremely important process in its own right, with many powerful applications.

When you differentiate x^n with respect to x you may think of the process involving two stages:

$x^n \longrightarrow$ multiply by the power $\longrightarrow$ reduce the power by 1 $\longrightarrow nx^{n-1}$

This *suggests* that the reverse process is given by:

$\dfrac{x^{n+1}}{n+1} \longleftarrow$ divide by the power $\longleftarrow$ increase the power by 1 $\longleftarrow x^n$

However, this doesn't give the *complete* picture. The reason is that you can differentiate x^n + (any constant) and still obtain nx^{n-1}. You need to take this into account when you reverse the process.

$\int$ is the symbol for integration and dx is used to show that the integration is with respect to the variable x.

KEY POINT

The result is written as $\int x^n \, dx = \dfrac{x^{n+1}}{n+1} + c$ where c is called the **constant of integration**. Note that $n \neq -1$.

Some examples are:

Notice that $\dfrac{1}{(\frac{3}{2})} = \dfrac{2}{3}$.

$$\int x^2 \, dx = \frac{x^3}{3} + c \qquad \int \sqrt{x} \, dx = \int x^{\frac{1}{2}} \, dx = \frac{2}{3}x^{\frac{3}{2}} + c \qquad \int 3 \, dx = 3x + c$$

$$\int \frac{1}{x^2} \, dx = \int x^{-2} \, dx = -x^{-1} + c = -\frac{1}{x} + c.$$

Sums and differences of functions are treated in the same way as in differentiation, by dealing with each term separately. The general rules are given below.

Function notation is useful here but you don't need to state the rules every time you use them.

KEY POINT

$$\int (f(x) \pm g(x)) \, dx = \int f(x) \, dx \pm \int g(x) \, dx$$

$\int af(x) \, dx = a \int f(x) \, dx$ (where a is a constant)

For example,

$$\int (3x^2 - 5x + 2) \, dx = x^3 - \tfrac{5}{2}x^2 + 2x + c.$$

In some situations you are given enough information to find the value of the constant.

Example Find the equation of the curve with gradient function $3x^2$ passing through (2, 5).

Using the point (2, 5) gives $5 = 2^3 + c$ so $c = -3$.

$$\frac{dy}{dx} = 3x^2 \Rightarrow y = \int 3x^2\,dx \Rightarrow y = x^3 + c. \quad \text{When } x = 2,\ y = 5 \Rightarrow c = -3.$$

The equation of the curve is $y = x^3 - 3$.

Progress check

1 Integrate each of these with respect to x.

(a) $\dfrac{1}{\sqrt{x}}$

(b) $\sqrt{x^3}$

2 A curve has gradient function $4x^3 + 1$ and passes through the point (1, 9). Find the equation of the curve.

1 (a) $2\sqrt{x} + c$ (b) $\frac{2}{5}x^{5/2} + c$

2 $y = x^4 + x + 7$

Sample questions and model answers

'Write down' means that you don't need to show any working to get the marks. But you might still find it useful as a way to clarify your thoughts.

Remember that $a^{-n} = \frac{1}{a^n}$

1

Write down the exact value of 3^{-3}.

(a) $3^{-3} = \frac{1}{3^3} = \frac{1}{27}$.

2

(a) Write $2x^2 - 12x + 11$ in the form $a(x+b)^2 + c$.

(b) State the minimum value of $2x^2 - 12x + 11$ and give the value of x where this occurs.

(c) Solve the equation $2x^2 - 12x + 11 = 0$ and express your answer in surd form.

(d) Sketch the graph of $y = 2x^2 - 12x + 11$.

Start by taking out the factor of 2 common to the first two terms.

Keep the 11 separate rather than have a fraction inside the brackets.

Complete the square and simplify.

The minimum value occurs when the expression in the brackets equals zero.

(a) $2x^2 - 12x + 11 = 2(x^2 - 6x) + 11$

$= 2((x-3)^2 - 9) + 11$

$= 2(x-3)^2 - 7$.

(b) The minimum value is -7 and this occurs when $x = 3$.

(c) $2x^2 - 12x + 11 = 0$

$\Rightarrow 2(x-3)^2 - 7 = 0$

$\Rightarrow (x-3)^2 = \frac{7}{2} = \frac{14}{4}$

$\Rightarrow x - 3 = \pm \frac{1}{2}\sqrt{14}$

$\Rightarrow x = 3 \pm \frac{1}{2}\sqrt{14}$.

(d)

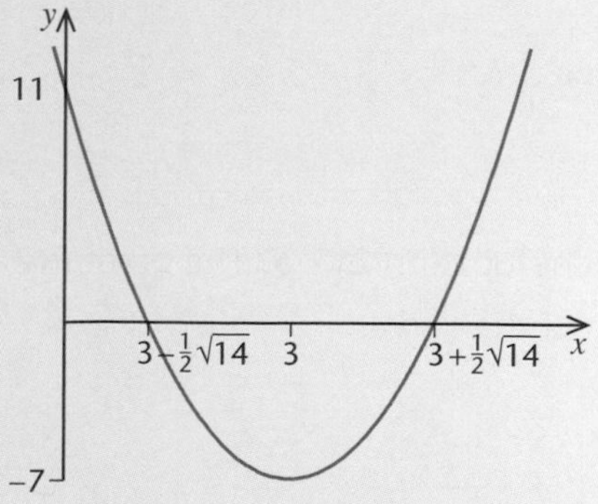

When $x = 0$, $y = 11$.

Use the information you have found to position the curve and label the key points.

Sample questions and model answers (continued)

3

A curve has equation $y = (2 - x)(2 + x)$.

The point P(−1, 3) lies on the curve.

(a) Find the gradient of the tangent to the curve at P.

(b) Find the equation of the normal to the curve at P in the form $px + qy = r$.

(c) The normal to the curve at P crosses the y-axis at A and the x-axis at B. Find the area of triangle AOB.

(a) $y = (2 - x)(2 + x)$

$= 4 - x^2$

Expand the brackets before differentiating.

$\frac{dy}{dx} = -2x.$

When $x = -1$, $\frac{dy}{dx} = -2(-1) = 2$

The gradient of the tangent at P is 2.

(b) The gradient of the normal at P is $-\frac{1}{2}$.

The product of the gradients must be −1, as they are at right angles to each other.

Equation of the normal at P:

$y - 3 = -\frac{1}{2}(x - (-1))$

This is using the equation of a straight line in the form $y - y_1 = m(x - x_1)$.

$y - 3 = -\frac{1}{2}(x + 1)$

$2(y - 3) = -(x + 1)$

$2y - 6 = -x - 1$

$x + 2y = 5$

Write the equation in the required form.

(c) When $x = 0$, $y = \frac{5}{2} \Rightarrow$ A is the point $(0, \frac{5}{2})$

When $y = 0$, $x = 5 \Rightarrow$ B is the point (5, 0)

Find where the line intersects the axes.

Area of triangle AOB $= \frac{1}{2} \times 5 \times \frac{5}{2} = \frac{25}{4} = 6\frac{1}{4}$ units^2.

4

A curve has gradient function $2x - 5$. The point P(2, −1) lies on the curve.

Find the equation of the curve.

$\frac{dy}{dx} = 2x - 5$

$y = \int (2x - 5)\,dx$

$= x^2 - 5x + c$

Remember the integration constant c here.

When $x = 2$, $y = -1$

$\Rightarrow -1 = 2^2 - 5(2) + c$

Use the fact that P lies on the curve to find the value of c.

$c = -1 - 4 + 10 = 5$

The curve has equation $y = x^2 - 5x + 5$.

Practice examination questions

1 Write down the exact value of $49^{-\frac{1}{2}}$.

2 Given that $8^{x-3} = 4^{x+1}$, find the value of x.

3 (a) Write $x^2 - 4x + 1$ in the form $(x + a)^2 + b$.

(b) Solve $x^2 - 4x + 1 = 0$ and express your answer in surd form.

(c) State the coordinates of the lowest point on the graph of $y = x^2 - 4x + 1$.

(d) Sketch the graph.

4 A function $f(x)$ is defined by:

$f(x) = x^3 + 4x^2 - 3x - 18$

for all real values of x.

(a) Sketch the graph of $y = f(x)$.

(b) Sketch the graphs of $y = f(x) + 10$ and $y = f(x - 2)$.

5 Find the coordinates of the points A and B at the intersection of these graphs.

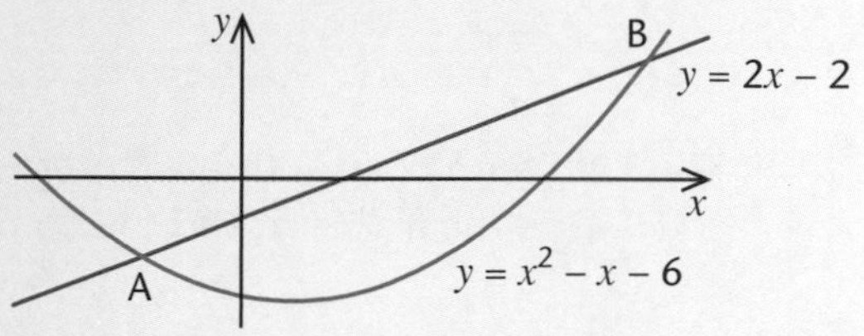

6 Solve the inequality

$3x^2 - 8x - 7 < 2x^2 - 3x - 11.$

Practice examination questions (continued)

7 (a) The diagram shows a straight line passing through the points A(–1,–5) and B(4, 1).
Find its equation in the form $ax + by + c = 0$.

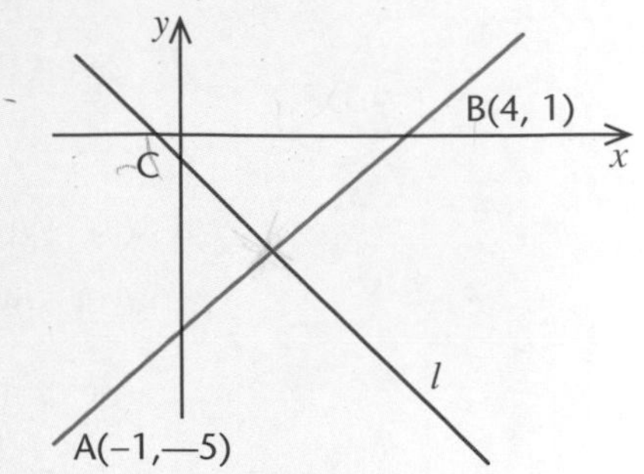

(b) A second line l, passes through the mid-point of A and B at right angles to AB.

Find its equation in the form $ax + by + c = 0$.

(c) The line l crosses the y-axis at C. Find the coordinates of C.

8 The diagram shows two lines l and m at right angles to each other. The equation of the line m is $3x + 2y = 6$. The line l passes through the point with coordinates (5, 4).
l and m cross the x-axis at A and B respectively.

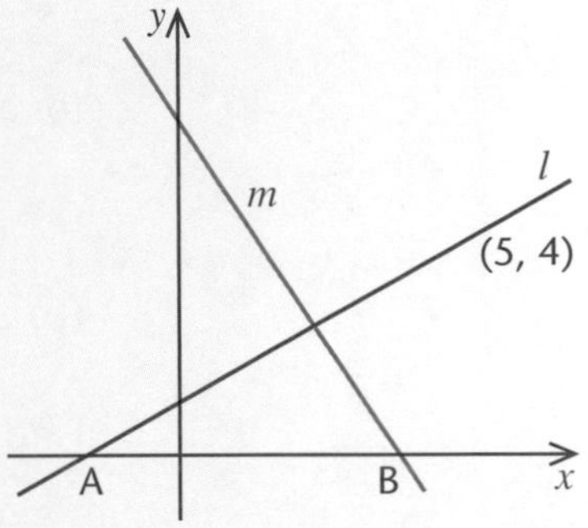

(a) Find the equation of the line l.

(b) Show that AB = 3.

9 The 10th term of an arithmetic progression is 74 and the sum of the first 20 terms is 1510. Find the first term and the common difference.

10 The diagram shows a straight line crossing the curve $y = (x - 1)^2 + 4$ at the points P and Q. It passes through (0, 7) and (7, 0).

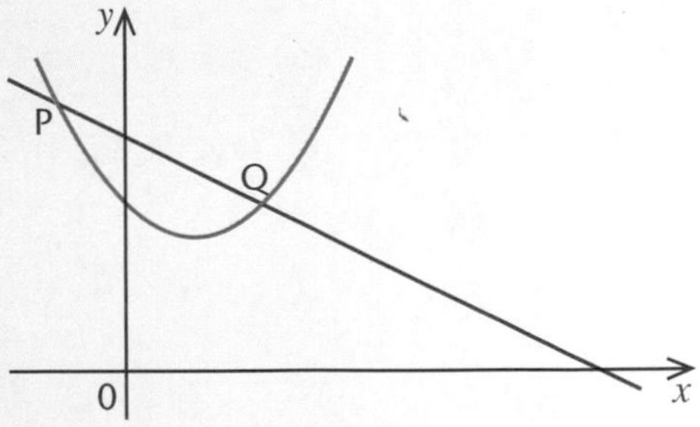

(a) Write down the equation of the straight line through P and Q.

(b) Find the coordinates of P and Q.

Chapter 2

Core 2 Pure Mathematics

The following topics are covered in this chapter:

- *Algebra and functions*
- *Coordinate geometry in the (x, y) plane*
- *Sequences and series*
- *Trigonometry*
- *Exponentials and logarithms*
- *Differentiation*
- *Integration*

2.1 Algebra and functions

After studying this section you should be able to:

- *carry out algebraic division*
- *use the remainder and factor theorems*

LEARNING SUMMARY

Algebraic division

If you work out 27 ÷ 4 for example, then you obtain 6 as the **quotient** and 3 as the **remainder**. You can write 27 = 4 × 6 + 3 to show the connection between these values.
The same thing applies in algebra when one polynomial is divided by another.

Example Find the quotient and remainder when $x^2 + 7x - 5$ is divided by $x - 4$.

Using $ax + b$ for the quotient and c for the remainder gives:

$$x^2 + 7x - 5 \equiv (x - 4)(ax + b) + c.$$
$$\equiv ax^2 + (b - 4a)x + c - 4b.$$

Equating coefficients of x^2 gives: $a = 1$.

Equating coefficients of x gives: $b - 4a = 7$

$\Rightarrow b - 4 = 7$ (using the result $a = 1$ from above)

$\Rightarrow b = 11$.

Equating the constant terms gives: $c - 4b = -5$

$\Rightarrow c - 44 = -5$ (using the result $b = 11$ from above)

$\Rightarrow c = 39$.

Substituting for a, b and c in [1] gives $x^2 + 7x - 5 \equiv (x - 4)(x + 11) + 39$.

The remainder theorem

In the previous example, the quotient and remainder of $(x^2 + 7x - 5) \div (x - 4)$ were found by using $x^2 + 7x - 5 \equiv (x - 4)(ax + b) + c$ and equating coefficients.

When $x = 4$, $(x - 4) = 0$ so $(x - 4)(ax + b) = 0$ and the RHS simplifies to c.

Another way to use this identity is to substitute particular values for x. Notice that when $x = 4$, the RHS simplifies to c and so the remainder is easily found by substituting $x = 4$ on the LHS. This gives $c = 4^2 + 7 \times 4 - 5 = 39$ as before.

In its generalised form this result is known as the **remainder theorem**, which states:

> When a polynomial $f(x)$ is divided by $(x - a)$ the remainder is $f(a)$.
>
> KEY POINT

Example Find the remainder when the polynomial $f(x) = x^3 + x - 5$ is divided by:

(a) $(x - 3)$ (b) $(x + 2)$ (c) $(2x - 1)$

Choose x so that $x - 3 = 0$.

Choose x so that $2x - 1 = 0$.

(a) The remainder is $f(3) = 3^3 + 3 - 5 = 25$

(b) The remainder is $f(-2) = (-2)^3 - 2 - 5 = -15$

(c) The remainder is $f(0.5) = 0.5^3 + 0.5 - 5 = -4.375$

The factor theorem

A special case of the remainder theorem occurs when the remainder is 0. This result is known as the **factor theorem**, which states:

> If $f(x)$ is a polynomial and $f(a) = 0$ then $(x - a)$ is a factor of $f(x)$.
>
> KEY POINT

Example Show that $(x + 3)$ is a factor of $x^3 + 5x^2 + 5x - 3$.

Taking $f(x) = x^3 + 5x^2 + 5x - 3$

$$f(-3) = (-3)^3 + 5(-3)^2 + 5(-3) - 3$$
$$= -27 + 45 - 15 - 3 = 0.$$

By the factor theorem $(x + 3)$ is a factor of $x^3 + 5x^2 + 5x - 3$.

Example A polynomial is given by $f(x) = 2x^3 + 13x^2 + 13x - 10$.

(a) Find the value of $f(2)$ and $f(-2)$.

(b) State one of the factors of $f(x)$.

(c) Factorise $f(x)$ completely.

(a) $f(2) = 2(2^3) + 13(2^2) + 13(2) - 10 = 84$

$f(-2) = 2(-2)^3 + 13(-2)^2 + 13(-2) - 10 = 0.$

(b) Since $f(-2) = 0$, $(x + 2)$ is a factor of $f(x)$ by the factor theorem.

$f(x)$ is a cubic so the linear factor $(x + 2)$ must be combined with a quadratic.

(c) $2x^3 + 13x^2 + 13x - 10 \equiv (x + 2)(ax^2 + bx + c)$.

This is an **identity** and so the values of a, b and c may be found by comparing coefficients.

Comparing the x^3 terms gives:	$a = 2$.
Comparing the x^2 terms gives:	$13 = 2a + b \Rightarrow b = 9$.
Comparing the constant terms gives:	$c = -5$.
It follows that:	$f(x) = (x + 2)(2x^2 + 9x - 5)$.

Factorising the quadratic part in the usual way gives $f(x) = (x + 2)(2x - 1)(x + 5)$.

Progress check

1 Find the remainder when $x^3 + 2x^2 - 5x + 1$ is divided by $(x - 2)$.

2 Given that $f(x) = 2x^3 + 7x^2 - 3x - 18$ and that $x = -2$ is a root of the equation $f(x) = 0$, factorise $f(x)$ completely.

1 7

2 $(x + 2)(x + 3)(2x - 3)$

2.2 Coordinate geometry in the (x, y) plane

After studying this section you should be able to:

- *use the equation of a circle and circle properties*

LEARNING SUMMARY

The equation of a circle

If the centre is at (0, 0), the equation is $x^2 + y^2 = r^2$

The equation of a circle with centre (a, b) and radius r is $(x - a)^2 + (y - b)^2 = r^2$.

KEY POINT

This result is based on Pythagoras' theorem.

$(x - a)^2 + (y - b)^2 = r^2$

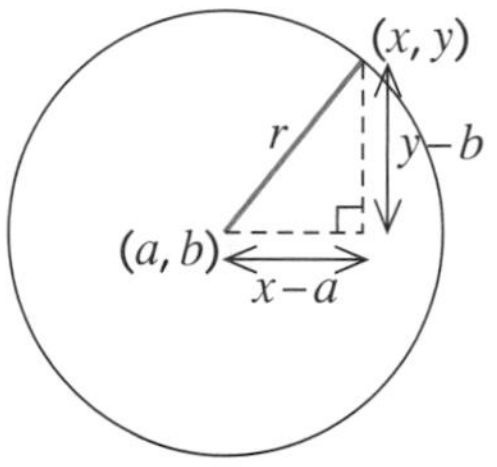

For example, the equation of a circle with centre (4, 0) and radius 5 units is $(x - 4)^2 + y^2 = 25$.

An alternative form of the equation of a circle is $x^2 + y^2 + 2gx + 2fy + c = 0$. The centre of the circle is $(-g, -f)$ and the radius is $\sqrt{g^2 + f^2 - c}$.

KEY POINT

For example, the circle with equation $x^2 + y^2 + 6x - 10y + 18 = 0$ has centre (–3, 5) and radius $\sqrt{(-3)^2 + 5^2 - 18} = \sqrt{16} = 4$ units.

You can also show this by completing the square on the x terms and on the y terms as follows:

$$x^2 + 6x + y^2 - 10y + 18 = 0$$
$$(x + 3)^2 - 9 + (y - 5)^2 - 25 + 18 = 0$$
$$(x + 3)^2 + (y - 5)^2 = 16$$

Compare with $(x - a)^2 + (y - b)^2 = r^2$.

The centre is at (–3, 5) and the radius is 4.

Circle properties

You need to remember and be able to use these **circle properties**.

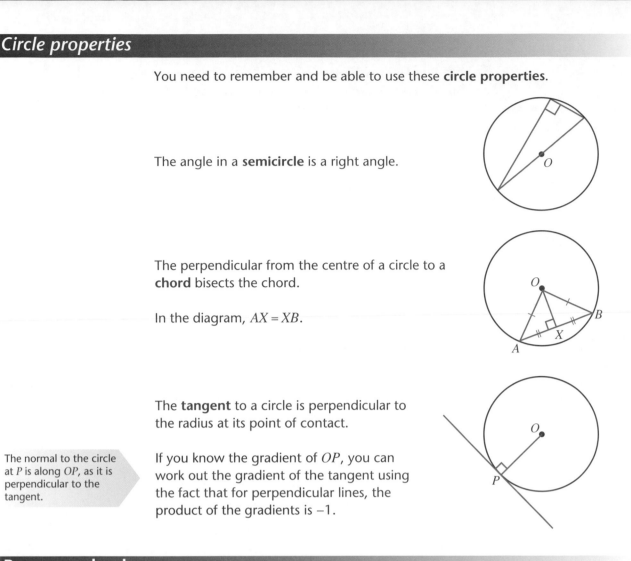

The angle in a **semicircle** is a right angle.

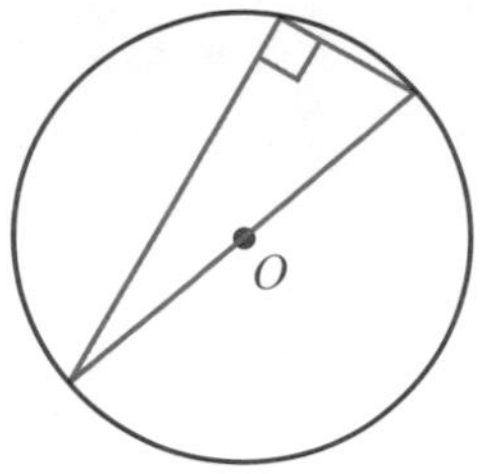

The perpendicular from the centre of a circle to a **chord** bisects the chord.

In the diagram, $AX = XB$.

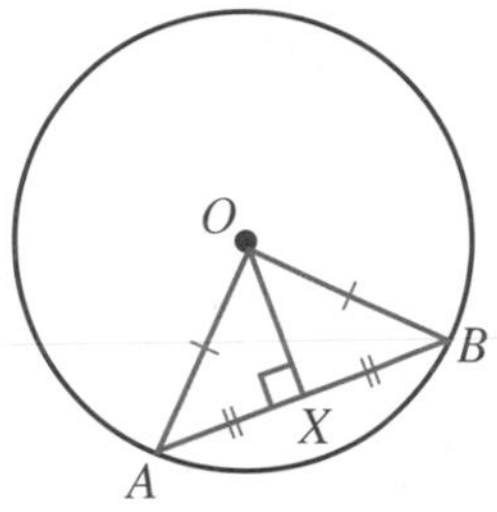

The **tangent** to a circle is perpendicular to the radius at its point of contact.

If you know the gradient of OP, you can work out the gradient of the tangent using the fact that for perpendicular lines, the product of the gradients is –1.

The normal to the circle at P is along OP, as it is perpendicular to the tangent.

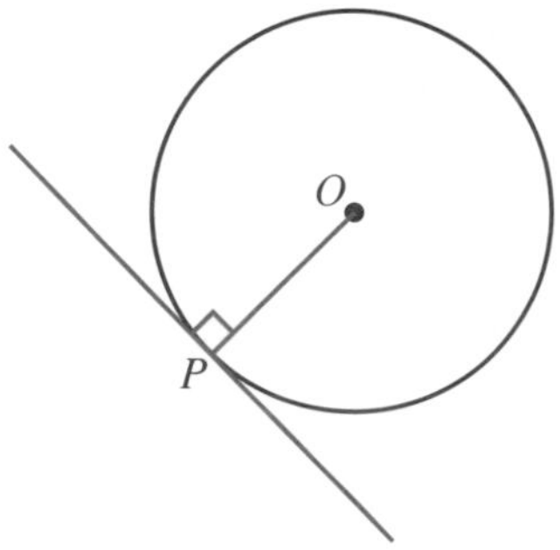

Progress check

1 (a) Find the equation of a circle with centre (–3, 8) and radius 5 units.
(b) A circle has equation $x^2 + y^2 - 12x + 8y + 43 = 0$.
Find its centre and radius.

1 (a) $(x+3)^2 + (y-8)^2 = 25$ (b) $(6, -4)$, 3 units

2.3 Sequences and series

After studying this section you should be able to:

- *understand the different types of sequence, including the notation and formulae used to describe them*
- *recognise geometric progressions and calculate the sum of their series*
- *determine whether a geometric series with an infinite number of terms has a finite sum and calculate its value when it exists*
- *use the binomial expansion for positive integer powers of n.*

LEARNING SUMMARY

Geometric series

Two special sequences are the **arithmetic progression** (A.P.), which was covered in Core 1, and the **geometric progression** (G.P.).

In a G.P. successive terms are connected by a **common ratio**, e.g. 3, 6, 12, 24, … . The definition of a G.P. may be given as $u_1 = a$, $u_{n+1} = ru_n$, where r is the common ratio.

The terms of a G.P. take the form $a, ar, ar^2, ar^3, \ldots$ and the nth term is given by $u_n = ar^{n-1}$.

A **series** is formed by adding together the terms of a sequence. The use of sigma notation can greatly simplify the way that series are written. For example, the series $1^2 + 2^2 + 3^2 + \ldots + n^2$ may be written as $\sum_{i=1}^{n} i^2$. The sum of the first n terms of a series is often denoted by S_n and so $S_n = u_1 + u_2 + u_3 + \ldots + u_n = \sum_{i=1}^{n} u_i$.

The sum of a G.P. is given by $S_n = a + ar + ar^2 + ar^3 + \ldots + ar^{n-1}$.
Multiplying throughout by r gives $rS_n = ar + ar^2 + ar^3 + \ldots + ar^{n-1} + ar^n$.
Subtracting gives: $S_n - rS_n = a - ar^n$

$$\Rightarrow S_n(1 - r) = a(1 - r^n)$$

You need to know how to establish this result.

So the sum of the first n terms of a geometric series is $S_n = \dfrac{a(1 - r^n)}{1 - r}$.

KEY POINT

If $r > 1$ then it can be convenient to use the result in the form $S_n = \dfrac{a(r^n - 1)}{r - 1}$.

Example Find the sum of the first 20 terms of the series $8 + 12 + 18 + 27 + \ldots$ to the nearest whole number.

In this series, $a = 8$, $r = 1.5$ and $n = 20$.

This gives $S_{20} = \dfrac{8(1.5^{20} - 1)}{1.5 - 1} = 53188.107\ldots = 53188$ to the nearest whole number.

> **KEY POINT**
> Provided that $|r| < 1$, the sum of a geometric series converges to $\frac{a}{1-r}$ as n tends to infinity. This is known as the **sum to infinity** of a geometric series.

For example, $1 + \frac{1}{2} + \frac{1}{4} + \frac{1}{8} + \ldots = \frac{1}{1 - \frac{1}{2}} = 2$.

Binomial expansion

An expression which has two terms, such as $a + b$ is called a **binomial**. The expansion of something of the form $(a + b)^n$ is called a **binomial expansion**. When n is a positive integer:

$$(a+b)^n = a^n + na^{n-1}b + \frac{n(n-1)}{2!}a^{n-2}b^2 + \frac{n(n-1)(n-2)}{3!}a^{n-3}b^3 + \ldots + b^n.$$

The situation where n is *not* a positive integer is dealt with in Core 4.

This expansion may appear complicated but it starts to look simpler if you follow the patterns from one term to the next:

Remember that $a^0 = 1$ when $a \neq 0$.

- Starting with a^n, the power of a is reduced by 1 each time until the last term, b^n, which is the same as a^0b^n.
- The power of b is increased by 1 each time. Notice that the first term, a^n, is the same as a^nb^0 and that the powers of a and b always add up to n in each term.
- It's worth remembering the first three coefficients: 1, n and $\frac{n(n-1)}{2!}$ then the rest follow the same pattern, giving $\frac{n(n-1)(n-2)}{3!}$, $\frac{n(n-1)(n-2)(n-3)}{4!}$ and so on.

Pascal's triangle: each row starts and ends with 1. Every other value is found by adding the pair of numbers immediately above it in the pattern.

```
  1   4   6
1   5   10
```

The coefficients in the expansion of $(a + b)^n$ also follow the pattern given by the row of **Pascal's triangle** that starts with 1 n …

```
            1
          1   1
        1   2   1
      1   3   3   1
    1   4   6   4   1
  1   5  10  10   5   1
```

So the coefficients in the expansion of $(a + b)^4$ for example are 1, 4, 6, 4 and 1.

It follows that
$(a+b)^4 = a^4 + 4a^3b + 6a^2b^2 + 4ab^3 + b^4$.

Variations on this result can be obtained by substituting different values for a and b. Some examples are:

In this case, $a = 1$ and $b = x$.

$$(1+x)^4 = 1 + 4x + 6x^2 + 4x^3 + x^4.$$

and

For this one, $a = 1$ and $b = -2x$.

$$\begin{aligned}(1-2x)^4 &= 1 + 4(-2x) + 6(-2x)^2 + 4(-2x)^3 + (-2x)^4 \\ &= 1 - 8x + 24x^2 - 32x^3 + 16x^4.\end{aligned}$$

The notation $\binom{n}{r}$ is often used to stand for the expression $\frac{n(n-1)\ldots(n-r+1)}{r!}$.

So, for example $\binom{n}{3} = \frac{n(n-1)(n-2)}{3!}$.

Using this notation, the binomial expansion may be written as:

$$(a+b)^n = a^n + \binom{n}{1}a^{n-1}b + \binom{n}{2}a^{n-2}b^2 + \binom{n}{3}a^{n-3}b^3 + \ldots + b^n.$$

- It's useful to recognise that the term involving b^r takes the form $\binom{n}{r}a^{n-r}b^r$.
- For positive integer values $\binom{n}{r}$ has the same value as ${}^nC_r$ and you may find that your calculator will work this out for you.
- When n is small and the full binomial expansion is required, the simplest way to find the coefficients is to use Pascal's triangle. For larger values of n it may be simpler to use the formula, particularly if only some of the terms are required.

Example Expand $(1 + 3x)^{10}$ in ascending powers of x up to and including the fourth term.

$$(1+3x)^{10} = 1 + \binom{10}{1}(3x) + \binom{10}{2}(3x)^2 + \binom{10}{3}(3x)^3 + \ldots$$
$$= 1 + 10(3x) + 45(9x^2) + 120(27x^3) + \ldots$$
$$= 1 + 30x + 405x^2 + 3240x^3 + \ldots.$$

Notice that the coefficient of x^7 is not totally determined by the value of $\binom{15}{7}$.

Example Find the coefficient of the x^7 term in the expansion of $(3 - 2x)^{15}$

The term involving x^7 is given by $\binom{15}{7}(3)^8(-2x)^7$

$$= 6435 \times 6561 \times -128x^7,$$

and so the required coefficient is –5 404 164 480.

Progress check

1 Find the sum of the first 20 terms of $12 + 15 + 18.75 + \ldots$ to 4 s.f.

2 Explain why the series $10 + 9 + 8.1 + 7.29 + \ldots$ is convergent and find the value of its sum to infinity to 5 s.f.

3 Expand $(1 + 3x)^{12}$ up to and including the third term.

1 4115
2 $|r| = 0.9 < 1$, 100.
3 $1 + 36x + 594x^2$

2.4 Trigonometry

LEARNING SUMMARY

After studying this section you should be able to:

- *state the exact values of the sine, cosine and tangent of special angles*
- *understand the properties of the sine, cosine and tangent functions*
- *sketch the graphs of trigonometric functions*
- *use the sine and cosine rules and the formula for the area of a triangle*
- *understand radian measure and use it to find arc lengths and areas of sectors*
- *solve trigonometric equations in a given interval*

Sine, cosine and tangent of special angles

The sine, cosine and tangent of 30°, 45°, and 60° may be expressed exactly as shown below. The results are found from Pythagoras' theorem.

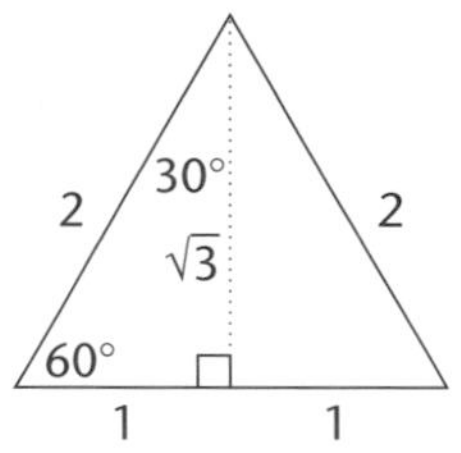

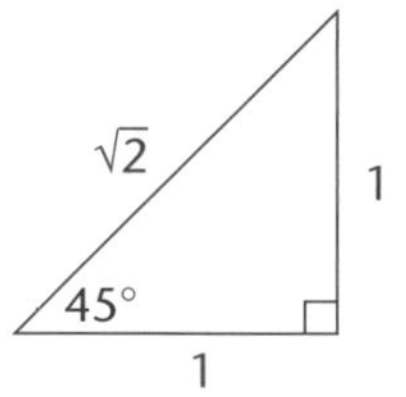

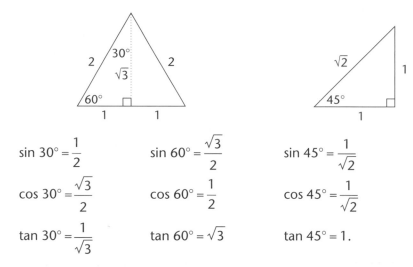

$\sin 30° = \frac{1}{2}$ $\quad$ $\sin 60° = \frac{\sqrt{3}}{2}$ $\quad$ $\sin 45° = \frac{1}{\sqrt{2}}$

$\cos 30° = \frac{\sqrt{3}}{2}$ $\quad$ $\cos 60° = \frac{1}{2}$ $\quad$ $\cos 45° = \frac{1}{\sqrt{2}}$

$\tan 30° = \frac{1}{\sqrt{3}}$ $\quad$ $\tan 60° = \sqrt{3}$ $\quad$ $\tan 45° = 1.$

Trigonometric graphs

You need to be able to sketch the graphs of the sine, cosine and tangent functions and to know their special properties.

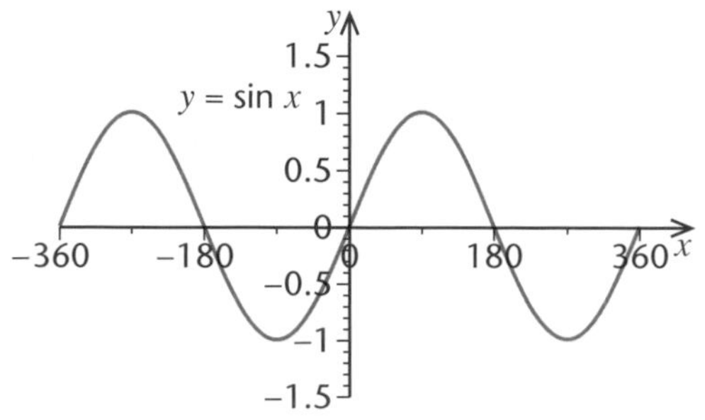

$\sin x$ is defined for any angle and always has a value between –1 and 1. It is a **periodic function** with period 360°.

The graph has **rotational symmetry** of order 2 about every point where it crosses the x-axis.

It has **line symmetry** about every vertical line passing through a vertex.

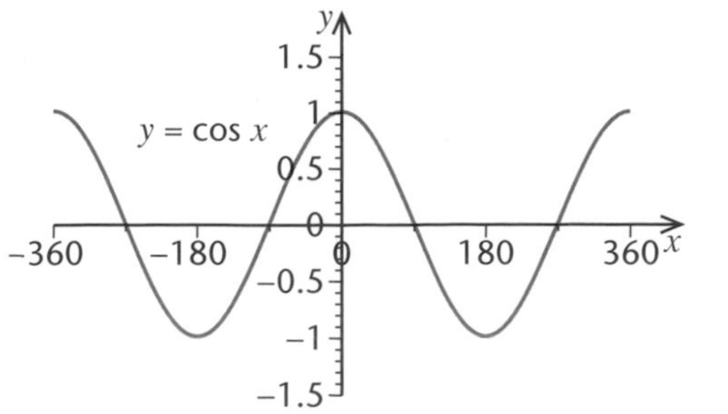

$\cos x \equiv \sin(x + 90°)$ so the graph of $y = \cos x$ can be obtained by translating the sine graph 90° to the left.

It follows that $\cos x$ is also a periodic function with period 360° and has the corresponding symmetry properties.

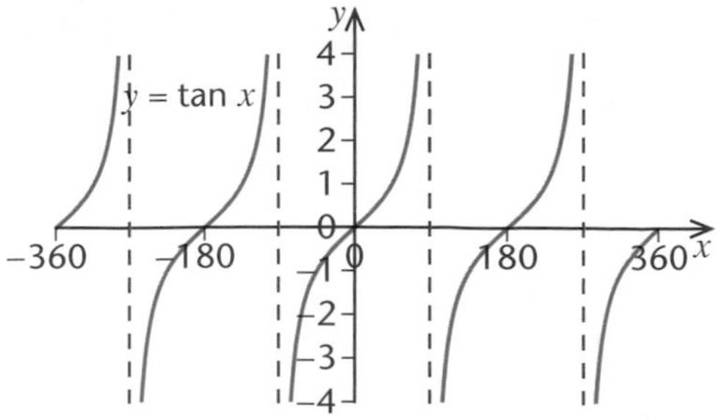

$\tan x \equiv \dfrac{\sin x}{\cos x}$.

$\tan x$ is undefined whenever $\cos x = 0$ and approaches $\pm\infty$ near these values. It is a periodic function with period 180°.

The graph has rotational symmetry of order 2 about 0°, ±90°, ±180°, ±270°, … .

The graphs of more complex trigonometric functions can often be produced by applying transformations to one of the basic graphs (see Core 1, page 18–19).

Example Sketch the graph of $y = 2\sin(x + 30°) + 1$ for $0° \leqslant x \leqslant 360°$.

It is helpful to think about building the transformations in stages.

Transformations			
Basic function	Translate the curve 30° to the left.	Now apply a one-way stretch with scale factor 2 parallel to the y-axis.	Finally translate the curve 1 unit up.
$y = \sin x$	$y = \sin(x + 30)$	$y = 2\sin(x + 30)$	$y = 2\sin(x + 30°) + 1$

The graph of $y = 2\sin(x + 30°) + 1$ may now be produced by applying these transformations in order starting from the graph of $y = \sin x$.

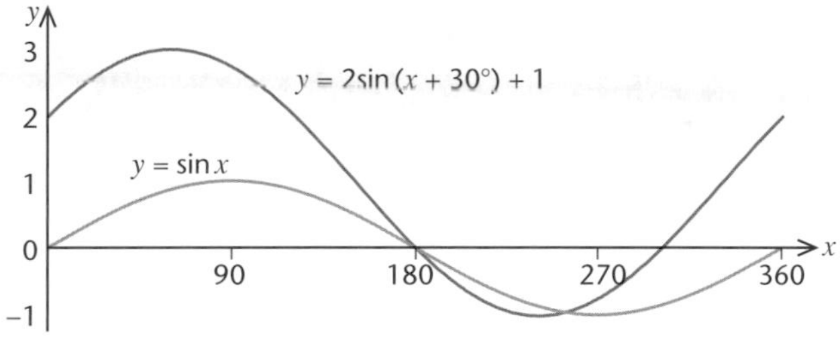

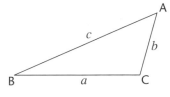

Area of a triangle

The same formula applies whether the angle is acute or obtuse.

Area of triangle $= \frac{1}{2}ab\sin C$

In both the diagrams, $b\sin C$ represents the height of the triangle.

The sine and cosine rules

A, B and C can be at any vertex. The opposite sides are then labelled as a, b, c.

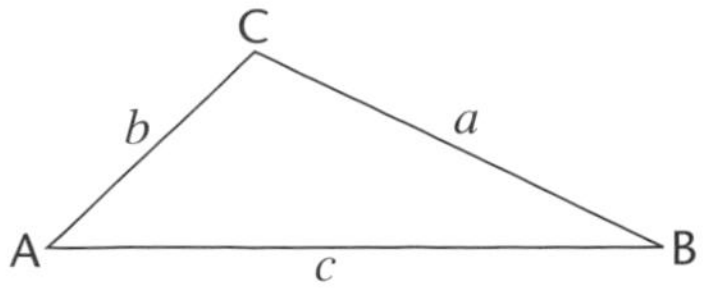

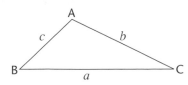

Sine rule: $\dfrac{a}{\sin A} = \dfrac{b}{\sin B} = \dfrac{c}{\sin C}$

Cosine rule: $a^2 = b^2 + c^2 - 2bc\cos A$

This may be rearranged to find an angle

$$\cos A = \frac{b^2 + c^2 - a^2}{2bc}$$

Radians

Angles can also be measured in **radians** and this makes it much easier to deal with trigonometric functions when using **calculus**.

1 radian $\approx 57.3°$. This may be written as $1^c \approx 57.3°$. However, the symbol for radians is not normally written when the angle involves π.
The following results are useful to remember:

$\pi = 180°, \frac{\pi}{2} = 90°, \frac{\pi}{4} = 45°, \frac{\pi}{3} = 60°, \frac{\pi}{6} = 30°.$

All the results that you know in degrees also apply in radians, e.g. $\sin\frac{\pi}{6} = 0.5$

Arc length and sector area

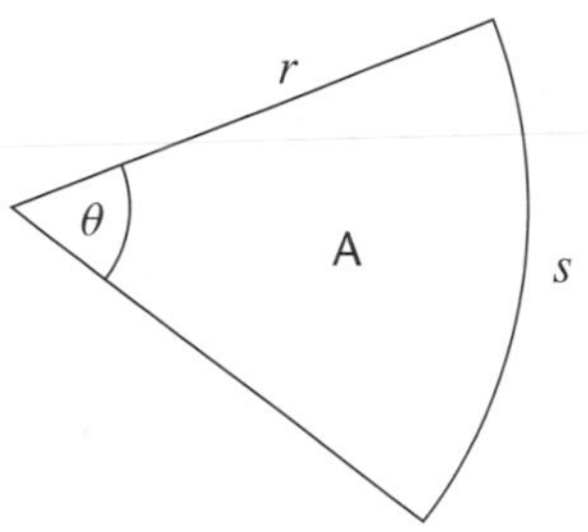

With θ in radians:

- Length of arc of sector is given by $s = r\theta$.
- Area of sector is given by $A = \frac{1}{2}r^2\theta$.

Solving trigonometric equations

You may be required to solve trigonometric equations in a variety of forms.

Example Solve $\sin(2x - 30°) = 0.6$ for $0° \leqslant x \leqslant 360°$.

First find a value of $2x - 30°$.

This is the principal value given by the calculator.

One value of $2x - 30°$ is $\sin^{-1}(0.6) = 36.86...°$.
You need to adjust the interval to find where the values of $2x - 30°$ must lie.

$$0° \leqslant x \leqslant 360° \Rightarrow 0° \leqslant 2x \leqslant 720° \Rightarrow -30° \leqslant 2x - 30° \leqslant 690°.$$

Now look at the graph of the sine function in this interval.

This is the graph of $y = \sin X$ for $-30° \leqslant X \leqslant 690°$ where $X = 2x - 30°$.
In other words, draw the sine curve as normal for the interval and then interpret the variable as $2x - 30°$.

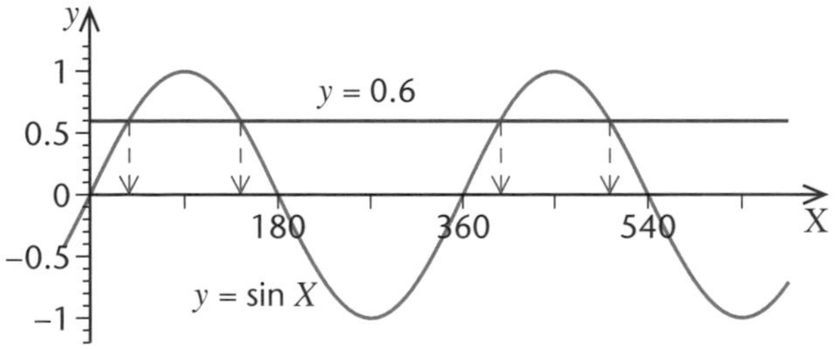

The graph shows that there are four values of $2x - 30°$ in the interval that make the equation work.

Using the symmetry of the curve, the values of $2x - 30°$ are:
$36.86...°, 180° - 36.86...°, 360° + 36.86...°, 540° - 36.86...°$.

This gives:

$2x - 30° = 36.86...° \Rightarrow x = 33.4°$
$2x - 30° = 143.14...° \Rightarrow x = 86.6°$

$2x - 30° = 396.86...° \Rightarrow x = 213.4°$
$2x - 30° = 503.14...° \Rightarrow x = 266.6°$
So $x = 33.4°, 86.6°, 213.4°, 266.6°$ (1 d.p.).

Using trigonometric identities

The following trigonometric relationships are **identities**. This means that they are true for all values of x.

They are very important when simplifying expressions and solving equations and should be learnt.

> $\tan x = \dfrac{\sin x}{\cos x}$
>
> KEY POINT

Example

Solve the equation $\sin x = \sqrt{3}\cos x$, for values of x between 0° and 360°.

This is allowed since $\cos x = 0$ is not a solution.

Divide each side by $\cos x$.

$$\frac{\sin x}{\cos x} = \sqrt{3}$$

$$\Rightarrow \tan x = \sqrt{3}$$

$$x = 60° \text{ or } 240°$$

The graph of $y = \tan x$ repeats every 180°, so to find the second solution, add 180°

> $\sin^2 x + \cos^2 x = 1$
>
> KEY POINT

This is sometimes known as the Pythagorean trigonometric identity.

Example Solve the equation $\sin^2 x - \cos^2 x - \sin x = 1$ for $-\pi \leqslant x \leqslant \pi$.

You need to recognise that this is a quadratic in $\sin x$.
It has to be rearranged in order to use the quadratic formula.

Replacing $\cos^2 x$ with $1 - \sin^2 x$ gives $\sin^2 x - (1 - \sin^2 x) - \sin x = 1$

$$\Rightarrow 2\sin^2 x - \sin x - 2 = 0.$$

$-1 \leqslant \sin x \leqslant 1$ for all values of x.

Using the formula gives $\sin x = \dfrac{1 \pm \sqrt{1 + 16}}{4} = \dfrac{1 \pm \sqrt{17}}{4}$

$$\Rightarrow \sin x = 1.28 \ldots \text{ (no solutions) or } \sin x = -0.78077 \ldots.$$

Remember to set your calculator to radians mode.

Using $\sin^{-1}$ gives $x = -0.8959^c$ to 4 d.p.

Another solution in the interval $-\pi \leqslant x \leqslant \pi$ is $-\pi + 0.8959^c$ giving $x = -2.2457^c$ to 4 d.p.

The diagram shows there are no other solutions in this interval.

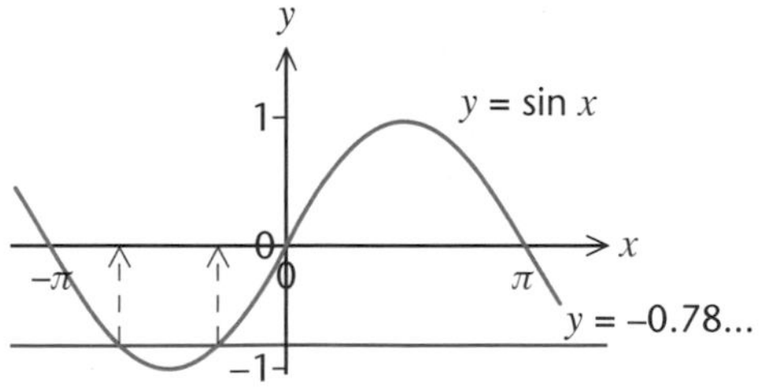

Progress check

1 Find the exact value of:

(a) sin 120° (b) cos 135° (c) tan 240°.

2 Describe the transformations needed to turn the graph of $y = \cos x$ into the graph of:

(a) $y = \cos 2x$ (b) $y = 3 \cos x$ (c) $y = 3 \cos(2x) - 1$.

3 Find the value of x and θ.

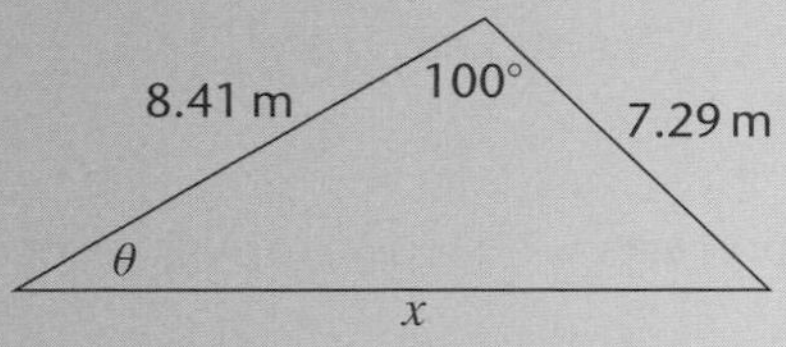

4 Find the exact value of:

(a) $\sin \frac{\pi}{4}$ (b) $\cos \frac{\pi}{3}$ (c) $\tan \frac{\pi}{6}$.

5 Solve $8 \sin x = 3 \cos x$ for $-360° \leqslant x \leqslant 360°$.

6 Solve $\cos(3x + 20°) = 0.6$ for $0° \leqslant x \leqslant 180°$.

7 Solve $2 \sin^2 x + \cos x = 1$ for $-\pi \leqslant x \leqslant \pi$.

8 OAB is a sector of a circle, centre O, radius 4 cm. Angle AOB = 0.5 radians.

(a) Find the perimeter of sector AOB.

(b) Find the length of the chord AB.

(c) Find the area of the sector OAB.

(d) Find the area of the triangle OAB.

(e) Find the area of the shaded segment.

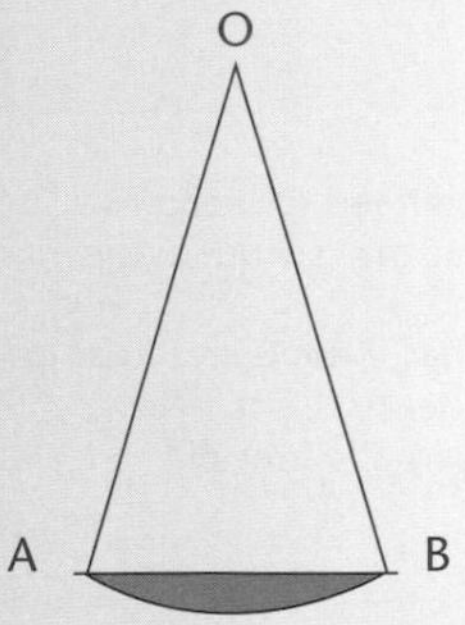

1 (a) $\frac{\sqrt{3}}{2}$ (b) $-\frac{1}{\sqrt{2}}$ (c) $\sqrt{3}$

2 (a) One-way stretch with scale factor 0.5 parallel to the x-axis.
(b) One-way stretch with scale factor 3 parallel to the y-axis.
(c) One-way stretch with scale factor 0.5 parallel to the x-axis, followed by a one-way stretch with scale factor 3 parallel to the y-axis, followed by a translation of 1 unit downwards.

3 $x = 12.0$ m (3 s.f.) $\theta = 36.6°$ (1 d.p.).

4 (a) $\frac{1}{\sqrt{2}}$ (b) $\frac{1}{2}$ (c) $\frac{1}{\sqrt{3}}$

5 $\tan x = \frac{3}{8}$ giving $x = 200.6°, 20.6°, -159.4°$ and $-339.4°$

6 11.0°, 95.6°, 131.1°

7 $\frac{2\pi}{3}, 0, -\frac{2\pi}{3}$

8 (a) 10 cm (b) 1.98 cm (2 d.p.) (c) 4 cm^2 (d) 3.84 cm^2 (2 d.p.) (e) 0.16 cm^2 (2 d.p.)

2.5 Exponentials and logarithms

After studying this section you should be able to:

- *understand exponential functions and sketch their graphs*
- *understand the log laws and use them to solve equations of the form $a^x = b$*

LEARNING SUMMARY

Exponential functions

An **exponential function** is one where the variable is a power or exponent. For example, any function of the form $f(x) = a^x$, where a is a constant, is an exponential function.

The diagram shows the graphs of $y = 2^x$ and $y = 10^x$.

Exponential functions are often used to represent, or model, patterns of:
- growth when $a > 1$
- decay when $a < 1$.

All graphs of the form $y = a^x$ cross the y-axis at (0, 1). Each graph has a different gradient at this point and the value of the gradient depends on a.

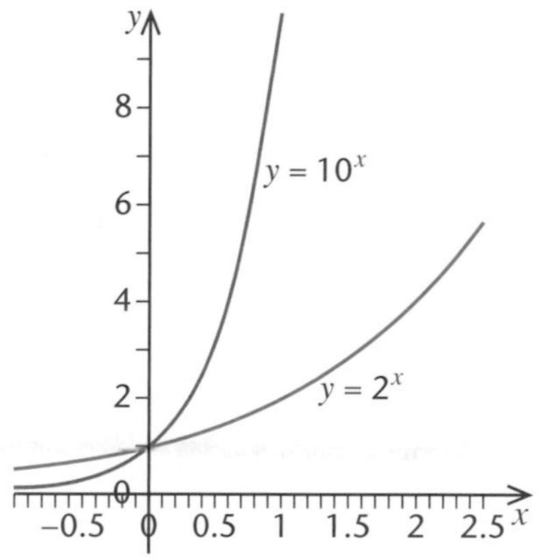

Logarithms

Exponential functions are related to **logarithmic functions** as follows:

$$y = \log_a x \Leftrightarrow x = a^y$$

$a > 0$ and $a \neq 1$.

a is called the base of the logarithm.

The **logarithm** (log) to a given **base** of a number is the **power** to which the base must be raised to equal the number.

Your calculator has a key for logs to the base 10, probably labelled log.

For example, $10^2 = 100$, so $\log_{10} 100 = 2$.

It is useful to remember the following hold for any base a:

$$\log_a a = 1, \text{ since } a^1 = a$$

$$\log_a 1 = 0, \text{ since } a^0 = 1.$$

The graph of $y = \log_a x$ is a reflection of $y = a^x$ in the line $y = x$.

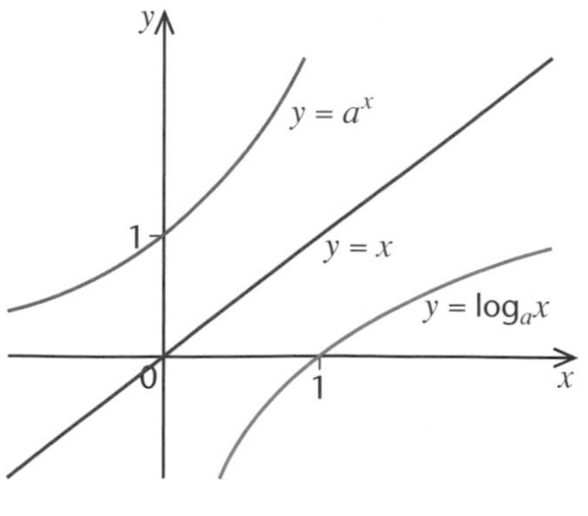

The log laws

There are three laws of logarithms that you need to know. The same laws apply in any base and so no particular base is stated:

- $\log a + \log b = \log ab$
- $\log a - \log b = \log\left(\frac{a}{b}\right)$
- $n \log a = \log a^n$.

You can use these laws to simplify expressions involving logs and to solve **exponential equations**, i.e. equations where the unknown value is a power.

Example Express $\log x + 3 \log y$ as a single logarithm.

$\log x + 3 \log y = \log x + \log y^3$ (using the third law)

$= \log xy^3$ (using the first law)

Example Solve the equation $5^x = 30$.

Taking logarithms of both sides gives:

$$\log 5^x = \log 30$$

$$\Rightarrow x \log 5 = \log 30$$

$$\Rightarrow x = \frac{\log 30}{\log 5} = 2.113 \text{ to 4 s.f}$$

This works in any base.

Change of the base of the logarithm

You can use the change of base formula to work out logarithms using any base.

$$\log_a b = \frac{\log_c b}{\log_c a}$$

So $\log_7 6 = \dfrac{\log_{10} 6}{\log_{10} 7} = 0.9208$

Progress check

1 Solve the equation $4^x = 100$ and give your answer to 4 s.f.

2 Write each of the following as a single logarithm.
(a) $\log_a x + 3 \log_a y - \frac{1}{2} \log_a z$
(b) $\log_{10} x - 1$

3 Evaluate the following:
(a) $\log_2 8$
(b) $\log_9 3$
(c) $2 \log_4 2$

1 $x = 3.322$
2 (a) $\log_a\left(\frac{xy^3}{\sqrt{z}}\right)$ (b) $\log_{10}\left(\frac{x}{10}\right)$
3 (a) 3 (b) $\frac{1}{2}$ (c) 1

2.6 Differentiation

After studying this section you should be able to:

- *recall differentiation methods and applications required in Core 1*
- *locate stationary points and use the second derivative method to distinguish between maxima and minima*
- *determine whether a function is increasing or decreasing in an interval*

LEARNING SUMMARY

Curve sketching

The gradient function gives information about the behaviour of a curve.

$f'(x) > 0$ at A and the function is **increasing**.

$f'(x) < 0$ at C and the function is **decreasing**.

$f'(x) = 0$ at B and D and the function is neither increasing nor decreasing. These points are called **stationary points**. B is at a **local maximum** and D is at a **local minimum**.

These types of stationary point are also known as **turning points**.

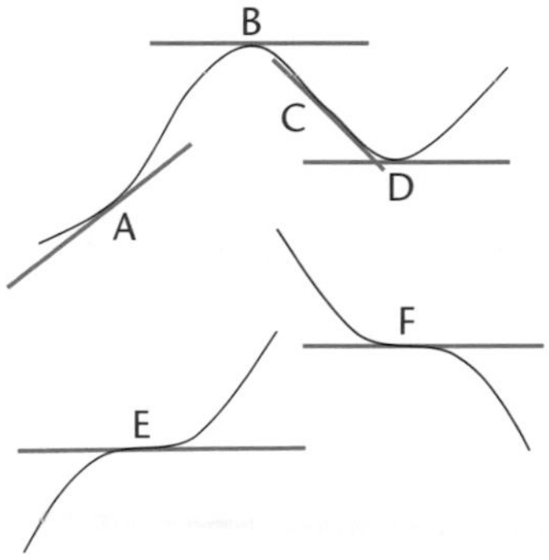

E and F show a different type of stationary point, called a **stationary point of inflexion**.

One way to distinguish between the types of stationary point is to look at the sign of the derivative on either side of the point.

- At a local maximum the sign of the derivative changes from positive to negative.
- At a local minimum the sign of the derivative changes from negative to positive.
- At a stationary point of inflexion there is no change of sign.

The second derivative

Starting from $\frac{dy}{dx} = f'(x)$ and differentiating again gives the **second derivative**. This is written as $\frac{d^2y}{dx^2} = f''(x)$. The second derivative can give information about the nature of any stationary points.

At a stationary point:

- $f''(x) > 0 \Rightarrow$ the point is a local minimum
- $f''(x) < 0 \Rightarrow$ the point is a local maximum.

But, if $f''(x) = 0$ then this gives no further information.

Example Find the stationary points of the curve $y = \frac{x^3}{3} - \frac{3}{2}x^2 + 2x - 1$ and use the second derivative to distinguish between them.

First differentiate the equation, then solve $\frac{dy}{dx} = 0$ to locate the stationary points.

$\frac{dy}{dx} = x^2 - 3x + 2.$

At a stationary point $\frac{dy}{dx} = 0 \Rightarrow x^2 - 3x + 2 = 0$

$$\Rightarrow (x-1)(x-2) = 0$$

$$\Rightarrow x = 1 \text{ or } x = 2.$$

When $x = 1$, $y = \frac{1}{3} - \frac{3}{2} + 2 - 1 = -\frac{1}{6}$.

When $x = 2$, $y = \frac{8}{3} - 6 + 4 - 1 = -\frac{1}{3}$.

The stationary points are $(1, -\frac{1}{6})$ and $(2, -\frac{1}{3})$.

In this case, the function is a cubic and so the shape of the curve is known. We should expect the first stationary point to be a local maximum and the second to be a local minimum.

Using the second derivative: $f''(x) = 2x - 3$

so $f''(1) = -1 < 0$ giving a local maximum,

and $f''(2) = 1 > 0$ giving a local minimum as expected.

Example This sheet of metal is bent to form an open box. Find the depth of the box which gives a maximum volume.

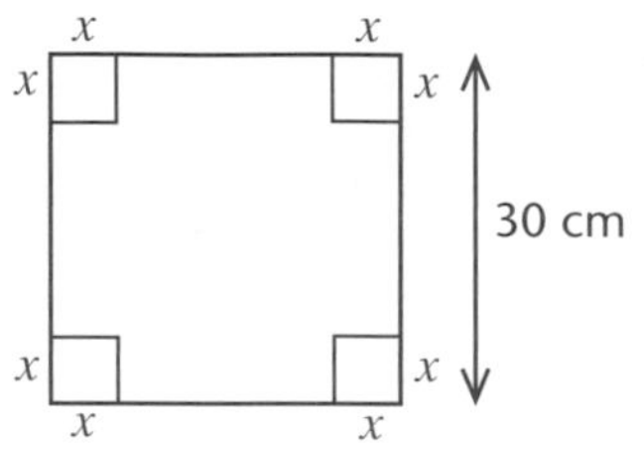

$V = x(30 - 2x)(30 - 2x) = 900x - 120x^2 + 4x^3$

$\frac{dV}{dx} = 900 - 240x + 12x^2 = 12(15 - x)(5 - x)$

$\frac{dV}{dx} = 0$ at turning points $x = 15$ and $x = 5$

$\frac{d^2V}{dx^2} = -240 + 24x$

$x = 15 \quad \frac{d^2V}{dx^2} = 120$ minimum

$x = 5 \quad \frac{d^2V}{dx^2} = -120$ maximum

Maximum volume at depth 5cm.

Progress check

1 Locate the stationary points on the curve $y = 2x^3 - 6x^2 - 18x + 5$ and use the second derivative to distinguish between them.

2 A rectangle of length x, where x varies, has a constant area of 48cm^2. Express the perimeter, y, in terms of x. Find the least possible value of y.

Solve the inequality $f'(x) < 0$.

3 Find the values of x for which the function $f(x) = x^3 - 6x^2 + 9x - 7$ is decreasing.

3 $1 < x < 3$

2 $y = 16\sqrt{3}$cm

1 Local maximum: $(-1, 15)$; local minimum $(3, -49)$.

2.7 Integration

After studying this section you should be able to:

- *evaluate a definite integral*
- *find the area under a curve*
- *use the trapezium rule to find an approximate value of an integral*

LEARNING SUMMARY

Indefinite integration

Indefinite integration was covered in Core 1; for revision, you should look at pages 31 to 32 and tackle the Progress check at the end of the section.

Example 1

$$\int (x^{\frac{3}{2}} + 2x^{-\frac{1}{2}})\,\mathrm{d}x = \frac{x^{\frac{5}{2}}}{\frac{5}{2}} + 2\frac{x^{\frac{1}{2}}}{\frac{1}{2}} + c$$

$$= \frac{2}{5}x^{\frac{5}{2}} + 4x^{\frac{1}{2}} + c$$

Example 2

Write the expression in index form first.

$$\int \frac{x+2}{\sqrt{x}}\,\mathrm{d}x = \int (x^{\frac{1}{2}} + 2x^{-\frac{1}{2}})\,\mathrm{d}x$$

$$= \frac{2}{3}x^{\frac{3}{2}} + 4x^{\frac{1}{2}} + c$$

Definite integration

An integral of the form $A = \int_a^b f(x)\mathrm{d}x$ is a **definite integral**.

It has a numerical value and is evaluated as follows:

Substitute the upper limit, then subtract the value when you substitute the lower limit.

$$\int_a^b f(x)\mathrm{d}x = [g(x)]_a^b = g(b) - g(a)$$

KEY POINT

A constant of integration is **not** needed in a definite integral.

For example $\int_1^4 2x\,\mathrm{d}x = [x^2]_1^4 = 4^2 - 1^2 = 15.$

Area under a curve

For areas *below* the x-axis, the definite integral gives a *negative* value.

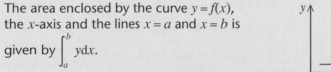

The area enclosed by the curve $y = f(x)$, the x-axis and the lines $x = a$ and $x = b$ is given by $\int_a^b y\mathrm{d}x$.

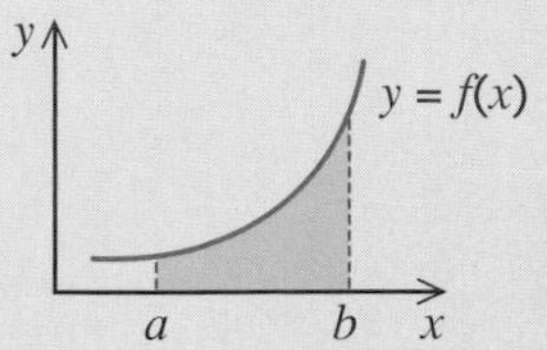

KEY POINT

Example Find the area under the curve $y = x^2 + 1$ between $x = 1$ and $x = 3$.

$$\text{Area} = \int_1^3 (x^2+1)\,dx = \left[\frac{x^3}{3}+x\right]_1^3$$
$$= \left(\frac{3^3}{3}+3\right)-\left(\frac{1^3}{3}+1\right)$$
$$= 9+3-\frac{1}{3}-1$$
$$= 10\tfrac{2}{3}.$$

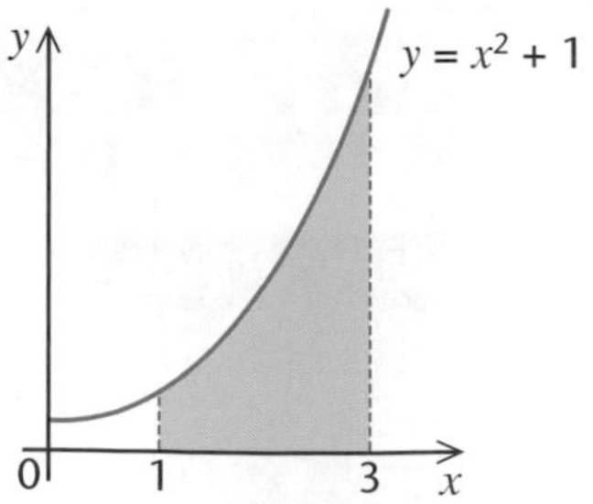

To find the **area between a line and a curve**, find the area under the line and the area under the curve separately and then subtract to find the required area. A sketch will be useful.

Progress check

1 Find the value of each of these definite integrals.

(a) $\int_0^5 (x-2)\,dx$ (b) $\int_4^9 \sqrt{x}\,dx$ (c) $\int_0^6 y^2\,dy$

2 Find the area under the curve $y = 3\sqrt{x}$ between $x = 1$ and $x = 9$.

2 52

1 (a) 2.5 (b) $12\frac{2}{3}$ (c) 72

Numerical integration – trapezium rule

The value of $A = \int_a^b y\,dx$ represents the area under the graph of $y = f(x)$ between $x = a$ and $x = b$. You can find an approximation to this value using the **trapezium rule**. This is especially useful if the function is difficult to integrate.

The area under the curve between a and b may be divided into n strips of equal width h.

It follows that $h = \dfrac{b-a}{n}$.

Each strip is approximately a trapezium and so the total area is approximately

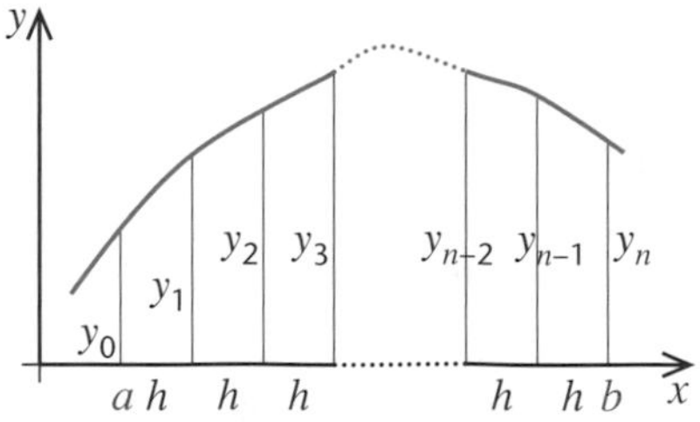

$$\tfrac{1}{2}h(y_0+y_1)+\tfrac{1}{2}h(y_1+y_2)+\tfrac{1}{2}h(y_2+y_3)+\ldots+\tfrac{1}{2}h(y_{n-2}+y_{n-1})+\tfrac{1}{2}h(y_{n-1}+y_n).$$

To increase the accuracy, use more strips.

KEY POINT

This simplifies to give the formula known as the **trapezium** rule:

$$\int_a^b y\,dx = \tfrac{1}{2}h\{(y_0+y_n)+2(y_1+y_2+\ldots+y_{n-1})\} \text{ where } h = \frac{b-a}{n}.$$

Example

Use the trapezium rule with five strips to estimate the value of $\int_1^2 2^x\,dx$.

It's useful to tabulate the information:

Work to a greater level of accuracy in your table than you intend to give in your final answer.

x	1	1.2	1.4	1.6	1.8	2
2^x	2	2.2974	2.6390	3.0314	3.4822	4
	y_0	y_1	y_2	y_3	y_4	y_5

$$h = \frac{2-1}{5} = 0.2$$

In this case, the working was done to 5 s.f. and the final answer is given to 3 s.f.

$$\text{So, } \int_1^2 2^x \, dx \approx 0.1\{(2+4) + 2(2.2974 + 2.6390 + 3.0314 + 3.4822)\}$$

$$= 2.89 \quad \text{to} \quad 3 \text{ s.f.}$$

Increasing gradient: overestimate

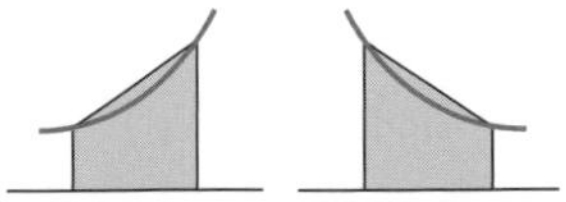

Decreasing gradient: underestimate

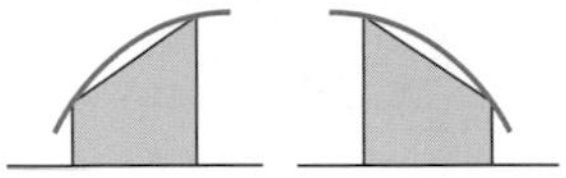

KEY POINT

If the gradient of the graph is increasing over the interval then the trapezium rule will give an over-estimate of the area. If the gradient of the graph is decreasing over the interval then it will give an under-estimate of the area.

A graphic calculator with a numerical integration function gives the value of the area as 2.8853901

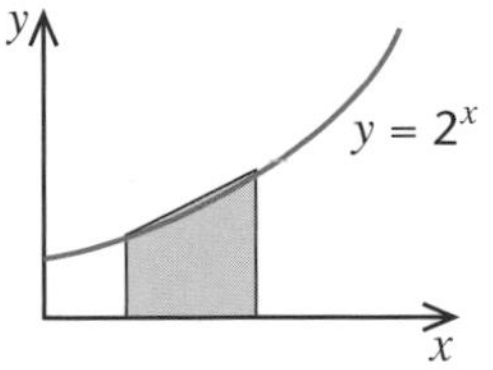

In this case, the gradient is increasing over the interval and so the trapezium rule gives an over-estimate of the area.

Progress check

1 Use the trapezium rule with five strips to estimate $\int_1^2 \frac{1}{x} \, dx$.

State whether the value obtained is an over-estimate or an under-estimate.

1 0.69563 (5 d.p.)
Over-estimate

Sample questions and model answers

1

Find the exact values of x, where $0 \leqslant x \leqslant 2\pi$, that satisfy the equation

$$\tan x = 2 \sin x.$$

Use the identity $\tan x = \dfrac{\sin x}{\cos x}$

$$\tan x = 2 \sin x$$

$$\frac{\sin x}{\cos x} = 2 \sin x$$

Multiply both sides by $\cos x$.

$$\sin x = 2 \sin x \cos x$$

$$\sin x - 2 \sin x \cos x = 0$$

Do not cancel $\sin x$ as this leads to loss of solutions.

$$\sin x(1 - 2 \cos x) = 0$$

Either $\sin x = 0 \Rightarrow x = 0, \pi, 2\pi$

or $1 - 2 \cos x = 0 \Rightarrow x = \dfrac{\pi}{3}, \dfrac{5\pi}{3}$

These are exact values.

So $x = 0, \dfrac{\pi}{3}, \pi, \dfrac{5\pi}{3}, 2\pi$

2

(a) Given that $\log_a x - \log_a 8 + 2 \log_a 4 = 0$, where a is a positive constant, find x.

(b) Find the value of x, correct to two decimal places, that satisfies the equation

$$3^{2x+1} = 40.$$

(a) $\log_a x - \log_a 8 + 2\log_a 4 = 0$

Simplify, using the log laws.

$\Rightarrow \quad \log_a x - \log_a 16 + \log_a 8 = 0$

$\Rightarrow \quad \log_a\left(\dfrac{16x}{8}\right) = 0$

$\Rightarrow \quad \log_a(2x) = 0$

Remember that $a^0 = 1$ for all values of a.

$\Rightarrow \quad 2x = 1$

$x = 0.5$

(b) $3^{2x+1} = 40$

Take logs to the base 10 of both sides.

$$\log_{10}(3^{2x+1}) = \log_{10}40$$

$$(2x + 1) \log_{10}3 = \log_{10}40$$

Use $\log_a(b^c) = c\log_a b$

$$2x + 1 = \frac{\log_{10}40}{\log_{10}3} = 3.3577\ldots$$

$$2x = 2.3577\ldots$$

$$x = 1.18 \text{ (2 d.p.)}$$

Sample questions and model answers (continued)

3

The third term of a geometric series is 20 and the fifth term is 5.

Given that the common ratio, r, is negative, find

(a) the value of r

(b) the first term,

(c) the sum to infinity of the series.

The nth term of a geometric series is ar^{n-1}.

(a) $ar^2 = 20$ (1)

$ar^4 = 5$ (2)

Dividing (2) by (1) gives $r^2 = \frac{1}{4}$

$r < 0 \Rightarrow r = -\frac{1}{2}$.

(b) Substitute into (1) $a(\frac{1}{4}) = 20$

$a = 80$

The series is convergent, since $|r| < 1$.

(c) $S_\infty = \dfrac{a}{1-r}$

$= \dfrac{80}{1-(-\frac{1}{2})}$

$= 53\frac{1}{3}$

4

(a) Express $\dfrac{5x^2-1}{\sqrt{x}}$ in the form $5x^a - x^b$, where a and b are rational numbers to be found.

(b) Hence find the exact value of $\displaystyle\int_1^2 \frac{5x^2-1}{\sqrt{x}}\,dx$.

(a) $\dfrac{5x^2-1}{\sqrt{x}} = \dfrac{5x^2}{x^{1/2}} - \dfrac{1}{x^{1/2}} = 5x^{3/2} - x^{-1/2}$

Remember that $\sqrt{x} = x^{1/2}$ and work in index form.

So $a = \frac{3}{2}$ and $b = -\frac{1}{2}$.

(b) $\displaystyle\int_1^2 \frac{5x^2-1}{\sqrt{x}}\,dx = \int_1^2 (5x^{3/2} - x^{-1/2})\,dx$

$= \left[\dfrac{5x^{5/2}}{\frac{5}{2}} - \dfrac{x^{1/2}}{\frac{1}{2}}\right]_1^2$

$= [2x^{5/2} - 2x^{1/2}]_1^2$

$(2)^{5/2} = (\sqrt{2})^5 = 4\sqrt{2}$

$= (8\sqrt{2} - 2\sqrt{2}) - (2-2)$

Leave the answer in surd form.

$= 6\sqrt{2}$

Practice examination questions

1 (a) Find the 8th term of a geometric progression with first term 3 and common ratio 2.

(b) Find the sum of the first 50 terms of the series $25 + 26.2 + 27.4 + \ldots$.

2 (a) Given that $(x + 2)$ is a factor of $f(x) = x^3 - 4x^2 - 3x + k$, use the factor theorem to find the value of k.

(b) Factorise $f(x)$ completely and sketch the graph of $y = f(x)$.

(c) Solve the inequality $f(x) \geqslant 0$.

(d) Find the remainder when $f(x)$ is divided by $(x + 1)$.

3 (a) Express $3 \log_a x - 2 \log_a y + \log_a(x + 1)$ as a single logarithm.

(b) Solve the equation $5^x = 100$. Give your answer to 3 d.p.

4 Solve the equation $\cos(2x + 30°) = 0.4$ for $0° \leqslant x \leqslant 360°$.

5 (a) Write the equation $\cos x + 3 \sin x \tan x - 2 = 0$ in the form $a \cos^2 x + b \cos x + c = 0$.

(b) Find the solutions of $\cos x + 3 \sin x \tan x - 2 = 0$ for $0° \leqslant x \leqslant 360°$.

6 Given that $f(x) = x^2 + \frac{1}{x}$, find

(a) $f'(2)$.

(b) $f''(-1)$.

7 The diagram shows a region R bounded by the curve $y = 3\sqrt{x}$, the x-axis and the lines $x = 1$ and $x = 4$.

Calculate the area of the region R.

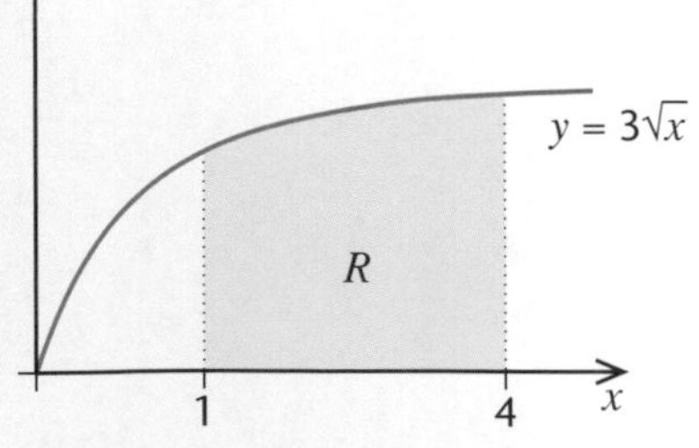

8 The diagram shows a circle of unit radius.

The angle between the radii is θ where $0 < \theta < \pi$ and the area of the shaded segment is $\frac{\pi}{5}$.

Show that $\theta = \sin\theta + \frac{2\pi}{5}$.

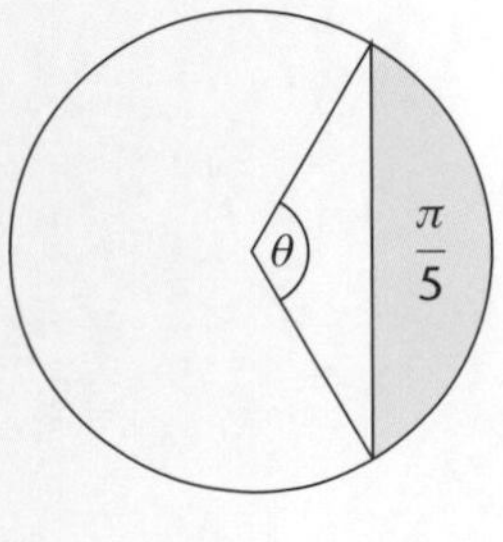

Practice examination questions (continued)

9 The area shaded in the diagram is bounded by the curve $y = \sqrt{x^2 - 3}$, the lines $x = 2$ and $x = 3$ and the x-axis.

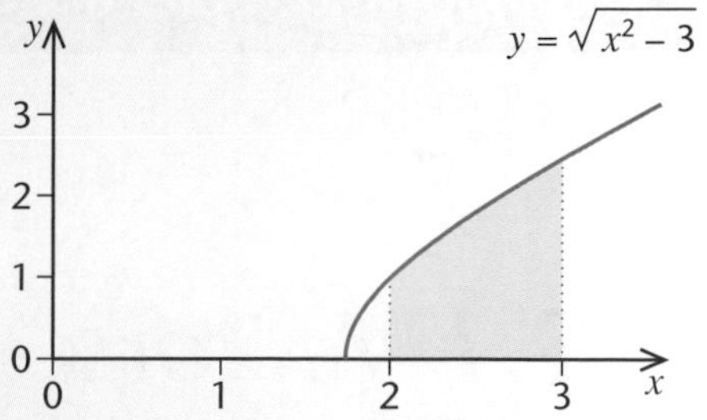

Express the shaded area as an integral and estimate its value using the trapezium rule with five intervals. Give your answer to 2 d.p.

10 (a) Factorise $2x^{\frac{3}{2}} - \dfrac{2x^{\frac{5}{2}}}{5}$.

(b) Find the points at which the graph of $y = 2x^{\frac{3}{2}} - \dfrac{2x^{\frac{5}{2}}}{5}$ meets the x-axis.

(c) Find $\dfrac{\mathrm{d}y}{\mathrm{d}x}$.

(d) Solve $\dfrac{\mathrm{d}y}{\mathrm{d}x} = 0$ for $x > 0$. Hence find the coordinates of the turning point of the graph, where x is positive, and determine its nature.

11 (a) Find the coordinates of the stationary points on the curve

$y = 2x^3 - 15x^2 - 36x + 10$.

(b) Use the second derivative to determine the nature of the stationary points found in (a).

(c) Find the set of values of x for which $2x^3 - 15x^2 - 36x + 10$ is a decreasing function of x.

Chapter 3

Mechanics 1

The following topics are covered in this chapter:

- *Vectors*
- *Kinematics*
- *Statics and moments*
- *Dynamics*

3.1 Vectors

LEARNING SUMMARY

After studying this section you should be able to:

- *understand the distinction between vector and scalar quantities*
- *add and subtract vector quantities and multiply by a scalar*
- *resolve vector quantities into two perpendicular components*
- *use the unit vectors* **i**, **j** *and* **k**
- *find the resultant, magnitude and direction of a vector*
- *understand the application of vectors in mechanics*

Vector and scalar quantities

A **scalar** quantity has size (or **magnitude**) but not direction. **Numbers** are scalars and some other important examples are **distance**, **speed**, **mass** and **time**.

A **vector** quantity has both size and **direction**. For example, distance in a specified direction is called **displacement**. Some other important examples are **velocity**, **acceleration**, **force** and **momentum**.

The diagram shows a **directed line segment**. It has size (in this case, length) and direction so it is a vector.

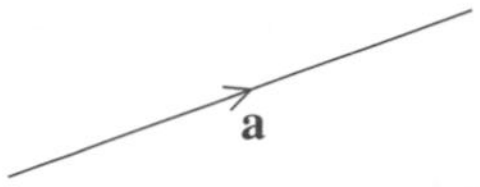

The diagram gives a useful way to represent *any* vector quantity and may also be used to represent addition and subtraction of vectors and multiplication of a vector by a scalar.

In a textbook, vectors are usually labelled with lower case letters in **bold** print.
When hand-written, these letters should be underlined e.g. $\underline{a}$

Addition and subtraction of vectors

This diagram shows three vectors **a**, **b** and **c** such that **c** = **a** + **b**.

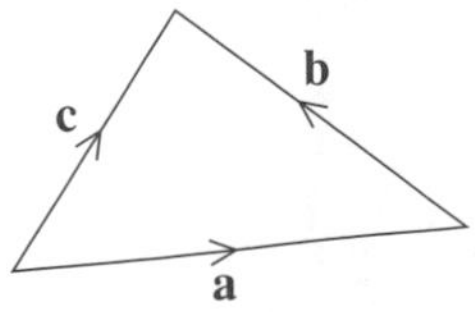

The vectors **a** and **b** follow on from each other and then **c** joins the start of vector **a** to the end of vector **b**.

Check that **a** + **b** = **b** + **a**.

The vector **c** is called the **resultant** of **a** and **b**.

On a vector diagram, −**q** has the opposite sense of direction to **q**.
Notice that **p** − **q** is represented as **p** + (−**q**).

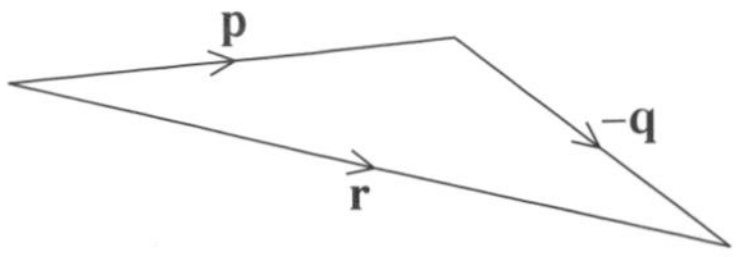

This diagram shows the same information in a different way.

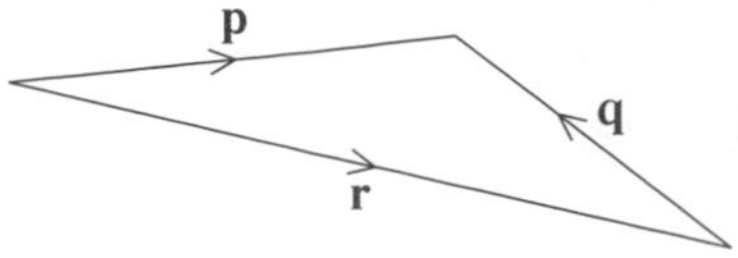

Following the route in the opposite direction to **q** is the same as adding −**q** or subtracting **q**.

Scalar multiplication

2**p** is parallel to **p** and has the same sense of direction but is twice as long.

−3**p** is parallel to **p** but has the opposite sense of direction and is three times as long.

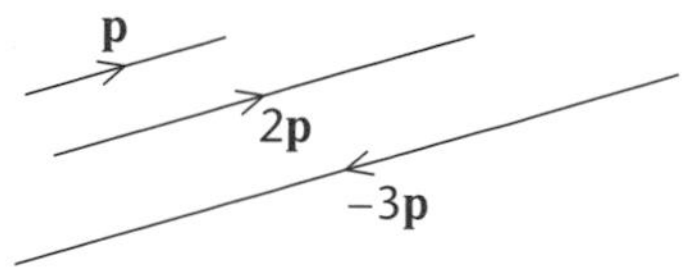

Component form

When working in two dimensions, it is often very useful to express a vector in terms of two special vectors **i** and **j**. These are **unit vectors** at right-angles to each other.
A vector **r** written as $\mathbf{r} = a\mathbf{i} + b\mathbf{j}$ is said to have **components** $a\mathbf{i}$ and $b\mathbf{j}$.

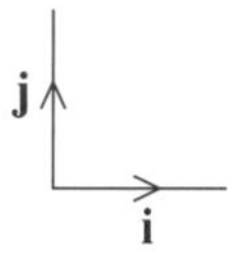

A unit vector is a vector of magnitude 1 unit.

Examples

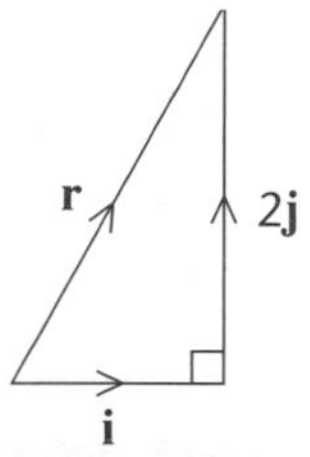

$\mathbf{r} = \mathbf{i} + 2\mathbf{j}$

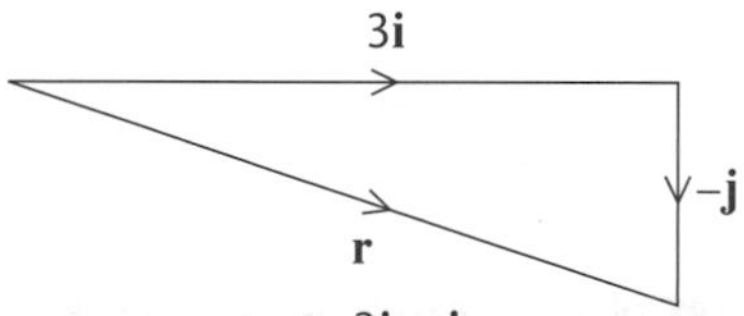

$\mathbf{r} = 3\mathbf{i} - \mathbf{j}$

For work involving the Cartesian coordinate system, **i** and **j** are taken to be in the positive directions of the x- and y-axes respectively.

In three dimensions, a third vector **k** is used to represent a unit vector in the positive direction of the z-axis.

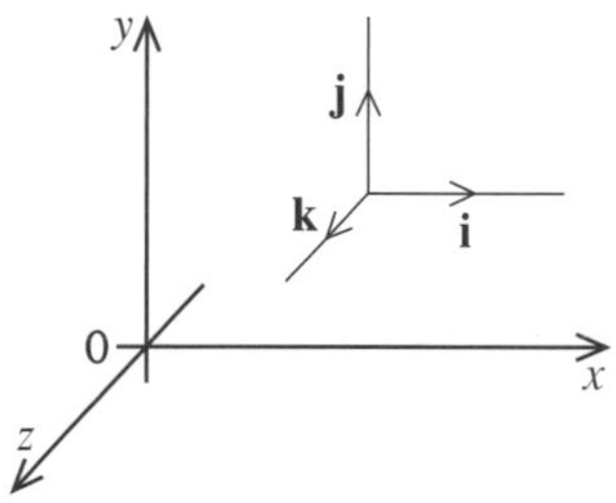

The **position vector** of a point P is the vector $\overrightarrow{OP}$ where O is the origin.

If P has coordinates (a, b) then its position vector is given by $\mathbf{r} = a\mathbf{i} + b\mathbf{j}$.

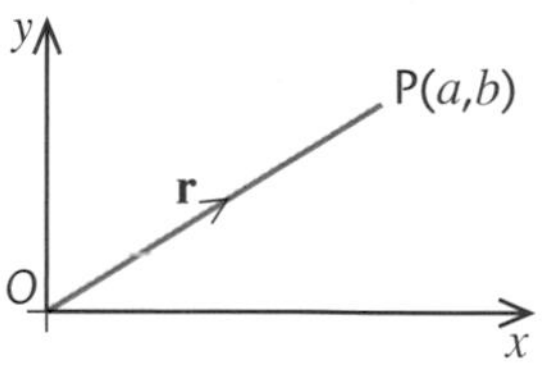

In three dimensions, a point with coordinates (a, b, c) would have position vector $\mathbf{r} = a\mathbf{i} + b\mathbf{j} + c\mathbf{k}$.

Resolving a vector

If you know the magnitude and direction of a vector then you can use trigonometry to **resolve** it in terms of **i** and **j** components.

The magnitude of **r** may be written as $|\mathbf{r}|$ or simply as r.

From the diagram, $\mathbf{r} = r\cos\theta\mathbf{i} + r\sin\theta\mathbf{j}$.

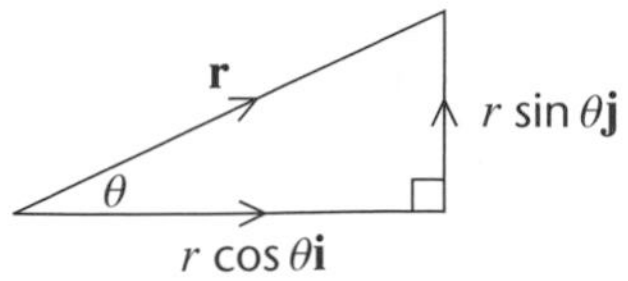

Example

A force **f** has magnitude 10 N at an angle of 60° above the horizontal. Express **f** in the form $a\mathbf{i} + b\mathbf{j}$ where **i** and **j** are unit vectors in the horizontal and vertical directions respectively.

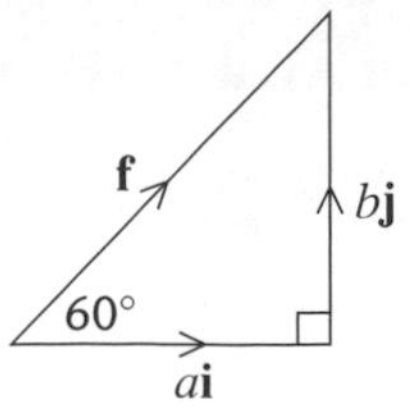

In this case, $a = 10 \cos 60° = 5$ and $b = 10 \sin 60° = 5\sqrt{3}$ giving $\mathbf{f} = 5\mathbf{i} + 5\sqrt{3}\mathbf{j}$.

Adding and subtracting vectors in component form

One advantage of expressing vectors in terms of **i** and **j** is that addition and subtraction may be done algebraically.

Example

$\mathbf{p} = 3\mathbf{i} + 2\mathbf{j}$ and $\mathbf{q} = 5\mathbf{i} - \mathbf{j}$. Express the following vectors in terms of **i** and **j**:

(a) $\mathbf{p} + \mathbf{q}$ (b) $\mathbf{p} - \mathbf{q}$ (c) $2\mathbf{p} - 3\mathbf{q}$.

(a) $\mathbf{p} + \mathbf{q} = (3\mathbf{i} + 2\mathbf{j}) + (5\mathbf{i} - \mathbf{j}) = 8\mathbf{i} + \mathbf{j}$

(b) $\mathbf{p} - \mathbf{q} = (3\mathbf{i} + 2\mathbf{j}) - (5\mathbf{i} - \mathbf{j}) = -2\mathbf{i} + 3\mathbf{j}$

(c) $2\mathbf{p} - 3\mathbf{q} = 2(3\mathbf{i} + 2\mathbf{j}) - 3(5\mathbf{i} - \mathbf{j}) = 6\mathbf{i} + 4\mathbf{j} - 15\mathbf{i} + 3\mathbf{j} = -9\mathbf{i} + 7\mathbf{j}$.

Finding the magnitude and direction of a vector

Another advantage of expressing a vector in terms of **i** and **j** components is that it is easy to find its magnitude and direction.

If $\mathbf{r} = a\mathbf{i} + b\mathbf{j}$ then the magnitude of **r** is given by

$$|\mathbf{r}| = \sqrt{a^2 + b^2}.$$

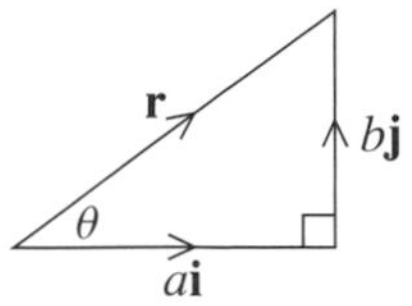

And the direction of ***r*** relative to ***i*** is given by

$$\theta = \tan^{-1}\left(\frac{b}{a}\right).$$

For example, if $\mathbf{r} = 3\mathbf{i} + 4\mathbf{j}$ then $|\mathbf{r}| = \sqrt{3^2 + 4^2} = 5$. and $\theta = \tan^{-1}(\tfrac{4}{3}) = 53.1°$.

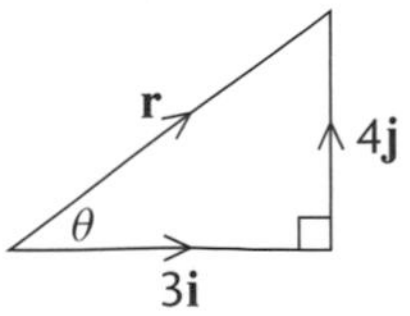

Progress check

1 Write down the position vector **r** of a point with coordinates (5, –2).

2

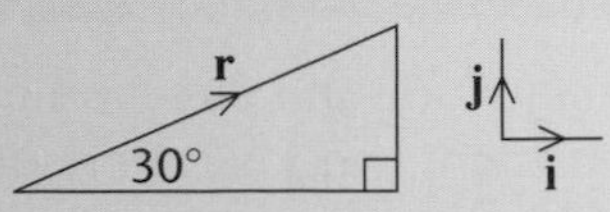

Given that $|\mathbf{r}| = 12$, express **r** in the form $\mathbf{r} = a\mathbf{i} + b\mathbf{j}$.

3 Find the magnitude of the vector $\mathbf{r} = 5\mathbf{i} - 12\mathbf{j}$ and give its direction relative to **i**.

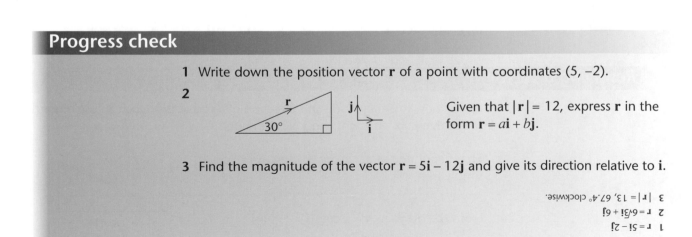

1 $\mathbf{r} = 5\mathbf{i} - 2\mathbf{j}$
2 $\mathbf{r} = 6\sqrt{3}\mathbf{i} + 6\mathbf{j}$
3 $|\mathbf{r}| = 13$, 67.4° clockwise.

3.2 Kinematics

After studying this section you should be able to:

- *apply the constant acceleration formulae for motion in a straight line*
- *draw and interpret graphs of displacement, velocity and acceleration against time*
- *use vectors to analyse motion in two and three dimensions*
- *apply the constant acceleration formulae to projectiles*

LEARNING SUMMARY

Motion in a straight line

You can use these formulae whenever the acceleration of an object is constant.

$v = u + at$	s is the displacement from a fixed position
$s = ut + \frac{1}{2}at^2$	u is the initial velocity
$s = \left(\frac{u+v}{2}\right)t$	a is the acceleration
	t is the time that the object has been in motion
$v^2 = u^2 + 2as$	v is the velocity at time t.

Example

An object starts from rest and moves in a straight line with constant acceleration 3 m s^{-2}. Find its velocity after 5 s.

From the given information: $u = 0$

$a = 3$

$t = 5$.

Don't include units in your working.

Using $v = u + at$

gives $v = 0 + 3 \times 5 = 15$.

Make a clear statement and include the appropriate units.

The velocity after 5 s is 15 m s^{-1}.

> Motion may take place in either direction along a straight line. One direction is taken to be positive for displacement, velocity and acceleration and the other direction is taken to be negative.
>
> KEY POINT

In some questions the acceleration due to gravity is taken to be 9.8 m s^{-2}.

Example

A stone is thrown vertically upwards with a velocity of 20 m s^{-1}. It has a downward acceleration due to gravity of 10 m s^{-2}.

(a) Find the greatest height reached by the stone.

(b) Find its velocity after 3 s.

(c) Find the height of the stone above the point of projection after 3 s.

(a) From the given information: $u = 20$

$a = -10$.

At the greatest height $v = 0$.

Using $v = u + at$

$0 = 20 - 10t \Rightarrow t = 2$.

Using $s = ut + \frac{1}{2}at^2$

gives $s = 20 \times 2 - 5 \times 4 = 20$

The greatest height reached is 20 m.

(b) Using

$v = u + at$ with $t = 3$

gives

$v = 20 - 10 \times 3 = -10$.

After 3 s the stone is moving downwards at 10 m s^{-1}.

(c) Using

$s = ut + \frac{1}{2}at^2$

gives

$s = 20 \times 3 - 5 \times 3^2 = 15$.

After 3 s the stone is 15 m above the point of projection.

Graphical representation

You need to know the properties of the graphs of distance, displacement, speed, velocity and acceleration against time.

KEY POINT

The gradient of a distance–time graph represents speed.
The gradient of a displacement–time graph represents velocity.
The gradient of a velocity–time graph represents acceleration.

The area under a speed–time graph represents the distance travelled.
The area under a velocity–time graph represents the change of displacement.
The area under an acceleration–time graph represents change in velocity.

Example

The diagram represents the progress of a car as it travels between two sets of traffic lights.
Find:

(a) The initial acceleration of the car.

(b) The deceleration of the car as it approaches the second set of lights.

(c) The distance between the traffic lights.

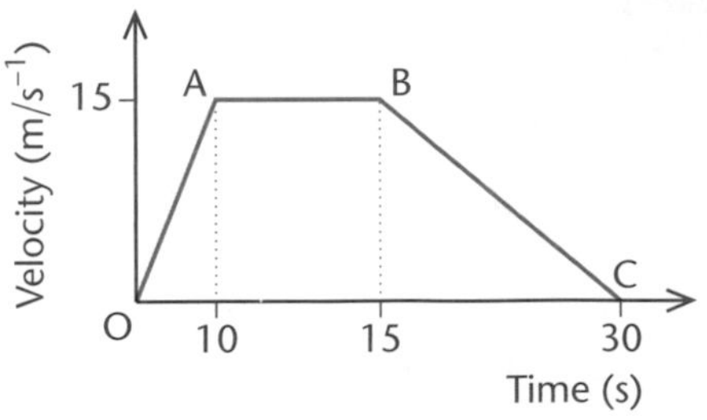

(a) Gradient of OA $= \frac{15}{10} = 1.5$
The initial acceleration of the car is 1.5 m s^{-2}.

(b) Gradient of BC $= -\frac{15}{15} = -1$.
The *acceleration* of the car is -1 m s^{-2} so the *deceleration* is 1 m s^{-2}.

(c) The area of the trapezium OABC is given by $\frac{15}{2}(5 + 30) = 262.5$,
so, the distance between the traffic lights is 262.5 m.

Using vectors

The constant acceleration formulae work equally well in two or three dimensions.

The work of this section may be extended to two and three dimensions using vectors.

Example

A particle has velocity $2\mathbf{i} - 3\mathbf{j}$ m s^{-1} when $t = 0$ and a constant acceleration of $\mathbf{i} + \mathbf{j}$ m s^{-2}. Find the speed of the particle when $t = 5$ and give its direction.

From the given information: $\mathbf{u} = 2\mathbf{i} - 3\mathbf{j}$

$\mathbf{a} = \mathbf{i} + \mathbf{j}$

$t = 5$.

Using $\mathbf{v} = \mathbf{u} + \mathbf{a}t$ gives $\mathbf{v} = 2\mathbf{i} - 3\mathbf{j} + 5(\mathbf{i} + \mathbf{j}) = 7\mathbf{i} + 2\mathbf{j}$.

Speed is the magnitude of the velocity, giving $v = \sqrt{7^2 + 2^2} = 7.28\ldots$.

So the speed of the particle when $t = 5$ is 7.28 m s^{-1} to 2 d.p.

From the diagram, $\tan\theta = \frac{2}{7} \Rightarrow \theta = 15.9°$.

When $t = 5$ the particle is moving at 15.9° anticlockwise to the direction of $\mathbf{i}$.

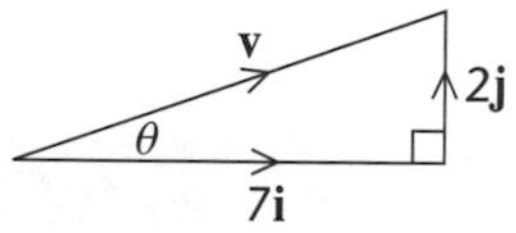

Progress check

1 A stone is thrown vertically upwards with a speed of 15 m s^{-1}. Take the downward acceleration due to gravity to be 10 m s^{-2} and find:
(a) The greatest height reached.
(b) The speed and direction of the stone after 2 s.

1 (a) 11.25 m (b) 5 m s^{-1} downwards

3.3 Statics and moments

After studying this section you should be able to:

- *resolve a single force into perpendicular components*
- *resolve a system of forces in a given direction*
- *find the resultant of a system of forces*
- *solve problems involving friction*
- *apply the conditions for equilibrium of coplanar forces in simple cases*
- *find the moment of a force*

LEARNING SUMMARY

Force

Force is a vector quantity that influences the motion of an object. It is measured in newtons (N). For example, the **weight** of an object is the force exerted on it by gravity. An object with a mass of m kg has weight mg N. **Tension**, **reaction** and **friction** are other examples of force and will be met in this section.

Resolving forces

Here are some examples of resolving forces into two components at right-angles to each other.
In each case the forces are represented in magnitude and direction by the sides of the triangle. The force to be resolved is *always* shown on the hypotenuse.

Notice the directions of the forces shown by the arrows. The diagram represents vector addition of the components.

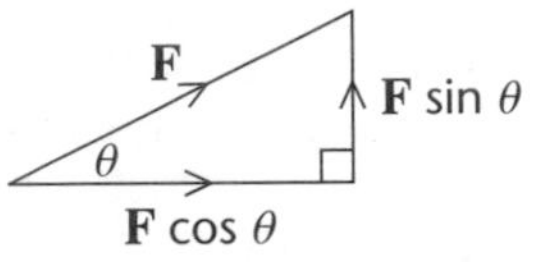

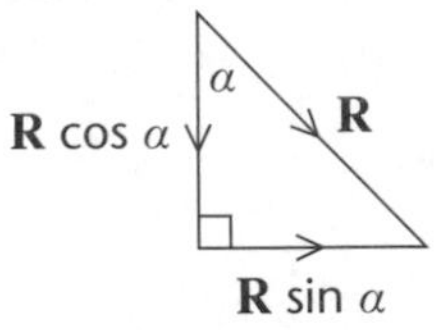

These are **vector diagrams** showing the relationship between a force and its components.

KEY POINT

Take care not to show both the force *and* its components on a **force diagram** or you will represent the force TWICE.

In a **force diagram**, you can *replace* a force with a pair of components that are equivalent to it. This will often make it easier to produce the equations necessary to solve a problem.

An important example is that of an object on an **inclined plane**.

Force diagram

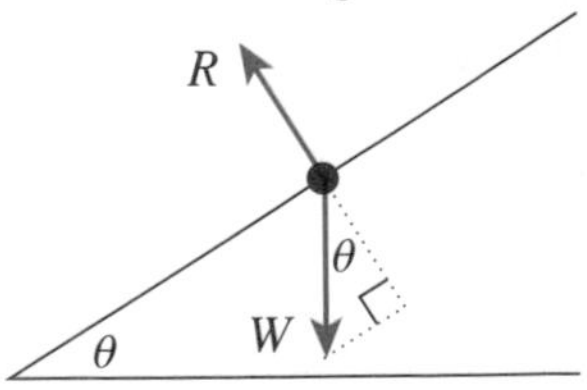

The diagram shows an object of weight W on a smooth plane inclined at angle θ to the horizontal. R is the force that the plane exerts on the object. It acts at right-angles to the plane and is called the **normal reaction**.

To analyse the behaviour of the object the weight is usually resolved into components parallel and perpendicular to the plane.

Vector diagram

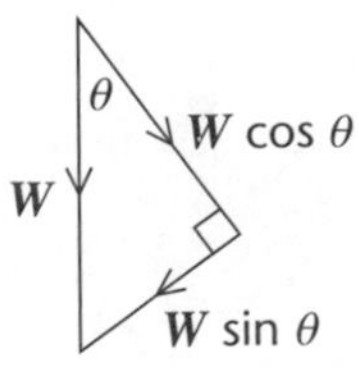

It's worth remembering these results.

This vector diagram shows the component of weight acting down the plane is $W \sin\theta$ and the component perpendicular to the plane is $W \cos\theta$.

Example

An object of weight 10 N is held in place on a plane inclined at 30° to the horizontal by two forces F N and R N as shown.

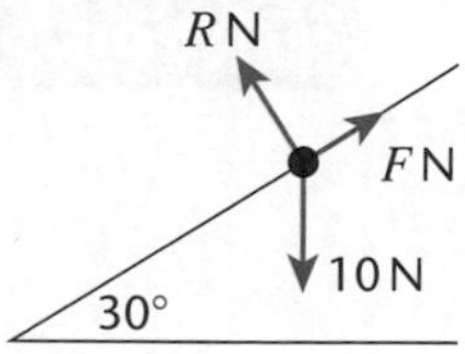

Find the values of F and R.

The forces F and R correspond to the components of the weight, parallel and perpendicular to the plane, but act in the opposite directions.

The forces must balance in each direction.

Resolving parallel to the plane: $F = 10 \sin 30° = 5$.

Resolving perpendicular to the plane: $R = 10 \cos 30° = 5\sqrt{3}$.

Resultant force

The effect of several forces acting at a point is the same as the effect of a single force called the **resultant force.**

You can find the resultant of a set of forces by using vector addition.

Example

Find the magnitude and direction of the resultant $\mathbf{R}$ N of the forces $(3\mathbf{i} - 2\mathbf{j})$ N, $(\mathbf{i} + 3\mathbf{j})$ N and $(-2\mathbf{i} + 4\mathbf{j})$ N.

The resultant force is given by: $\mathbf{R} = (3\mathbf{i} - 2\mathbf{j}) + (\mathbf{i} + 3\mathbf{j}) + (-2\mathbf{i} + 4\mathbf{j})$

$= 2\mathbf{i} + 5\mathbf{j}$.

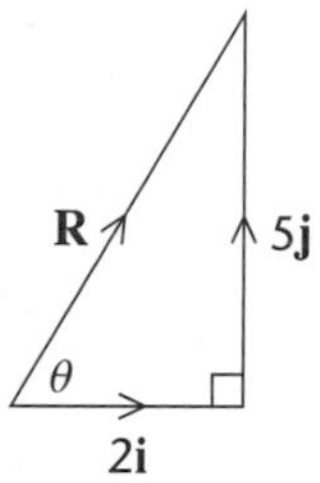

The magnitude of the resultant is $|\mathbf{R}| = \sqrt{2^2 + 5^2} = \sqrt{29}$.

From the diagram: $\theta = \tan^{-1}(\frac{5}{2}) = 68.2°$

The resultant has magnitude $\sqrt{29}$ N and acts at 68.2° anticlockwise from the direction of $\mathbf{i}$.

Equilibrium

A set of forces acting at a point is in **equilibrium** if the resultant force is zero.

It follows that the resolved parts of the forces must balance in any chosen direction.

Example

Find the force $\mathbf{F}$ such that the forces $(5\mathbf{i} - 3\mathbf{j})$, $(2\mathbf{i} + \mathbf{j})$ and $\mathbf{F}$ are in equilibrium.

For equilibrium $(5\mathbf{i} - 3\mathbf{j}) + (2\mathbf{i} + \mathbf{j}) + \mathbf{F} = 0$

$\Rightarrow 7\mathbf{i} - 2\mathbf{j} + \mathbf{F} = 0$

$\Rightarrow \mathbf{F} = -7\mathbf{i} + 2\mathbf{j}$.

Friction

Whenever two rough surfaces are in contact, the tendency of either surface to move relative to the other is opposed by the **force of friction**.

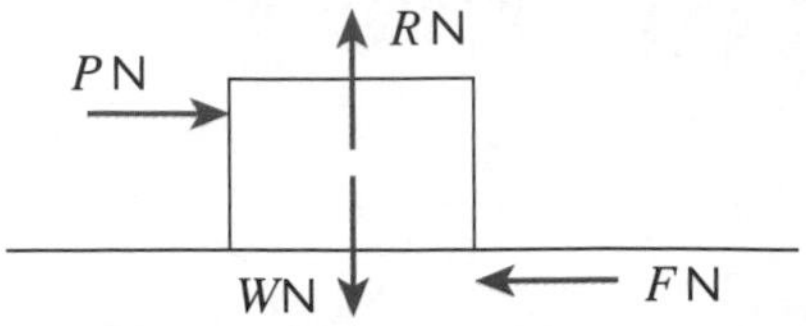

The diagram shows an object of weight W N resting on a rough horizontal surface. The object is pushed from one side by a force P N and friction responds with force F N in the opposite direction.

For small values of P, no movement takes place and $F = P$.

If P increases then F increases to maintain equilibrium until F reaches a maximum value. At this point the object is in **limiting equilibrium** and is on the point of slipping.

If P is now increased again then F will remain the same. The equilibrium will be broken and the object will move.

The maximum value of F depends on:

- The magnitude of the normal reaction R.
- The roughness of the two surfaces measured by the value μ. This value is known as the **coefficient of friction**.

In general: $F \leqslant \mu R$.
For limiting equilibrium: $F = \mu R$.

Example
An object of mass 8 kg rests on a rough horizontal surface. The coefficient of friction is 0.3 and a horizontal force P N acts on the object which is about to slide. Take $g = 10\ \text{m s}^{-2}$ and find the value of P.

Remember that the weight of the object is given by mg.

Resolving vertically gives $R = 80$.

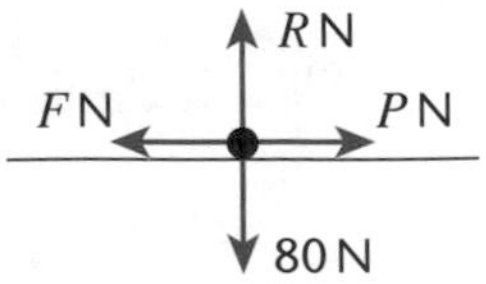

Friction is limiting so $F = \mu R = 0.3 \times 80 = 24$.

Resolving horizontally: $P = F \Rightarrow P = 24$.

Example
In the diagram, the object of mass m kg is on the point of slipping down the plane. The coefficient of friction between the object and the plane is μ. Show that $\mu = \tan\theta$.

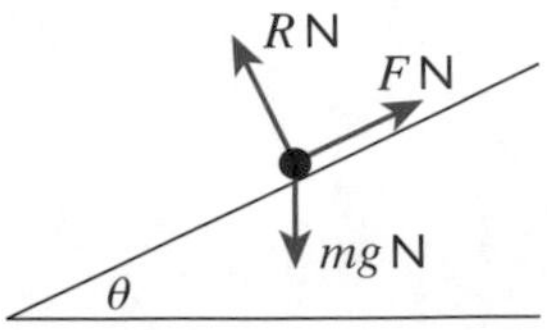

Resolving perpendicular to the plane $R = mg\cos\theta$.
Resolving parallel to the plane $F = mg\sin\theta$.
Friction is limiting so $F = \mu R$.

This gives $mg\sin\theta = \mu mg\cos\theta \Rightarrow \mu = \dfrac{\sin\theta}{\cos\theta} = \tan\theta$.

Moments

The **moment of a force** about a point is a measure of the turning effect of the force about the point.
It is found by multiplying the force by its perpendicular distance from the point.
The moment of a force acts either clockwise ↻ or anticlockwise ↺ about the point.

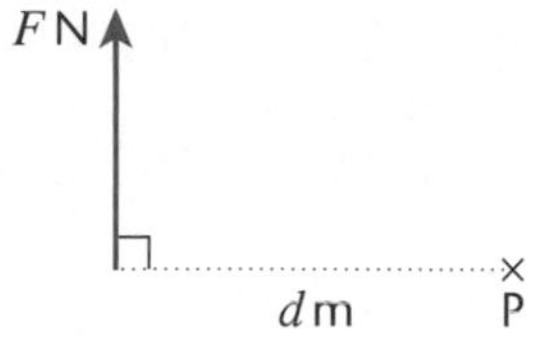

The moment of F about P is Fd N m ↻.

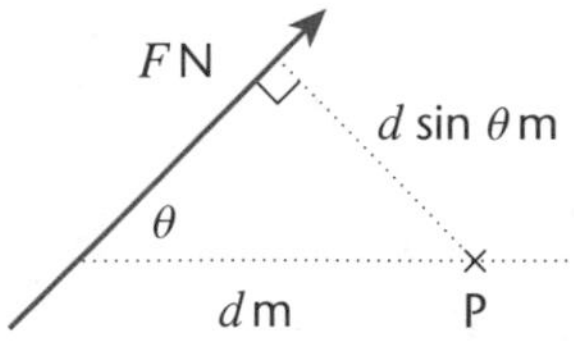

The moment of F about P is $Fd \sin\theta$ N m ↻.

The units of force (N) are multiplied by the units of distance (m) to give N m.

When there is more than one force, the **resultant moment** about a point is found by adding the separate moments. One direction is taken to be positive and the other negative.

Example Find the resultant moment about O of the forces shown in the diagram.

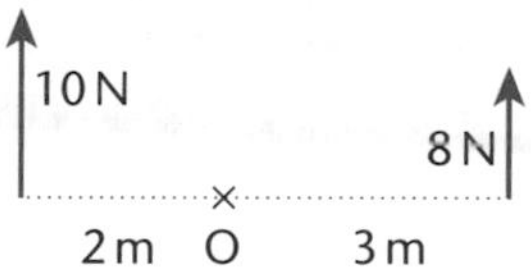

Taking ↺ as positive:
The total moment is $8 \times 3 - 10 \times 2$ N m ↺
$= 4$ N m ↺.

Equilibrium

For an object to be in equilibrium, the resultant force acting on it must be zero and any resultant moment must be zero.

In this context, a light rod is a rod is taken to have a mass that is so small that it can be ignored in any calculations.

Example
The diagram shows a light rod AB in equilibrium. Find the values of F and d.

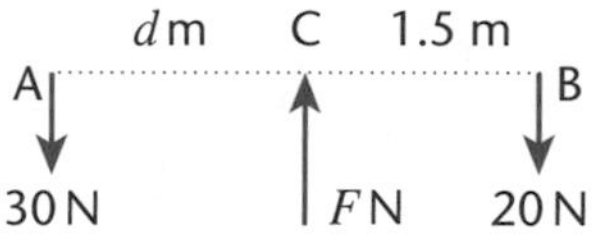

Resolving vertically $F - 30 - 20 = 0$
$\Rightarrow$ $F = 50$

Taking moments about C (written as M(C)) will give an equation involving d.

M(C) gives $30d - 20 \times 1.5 = 0$
$\Rightarrow$ $30d = 30$
$\Rightarrow$ $d = 1$.

Example
If F_1 = 10N and F_2 = 15N, find F_3 and θ.

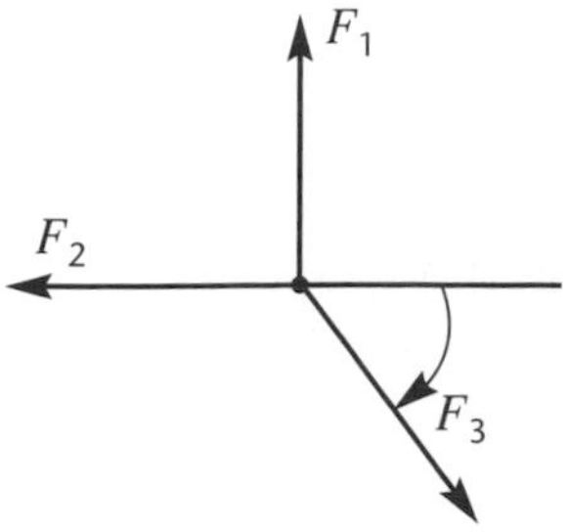

Resolving ↔ $F_2 = 15 = F_3 \cos\theta$
↕ $F_1 = 10 = F_3 \sin\theta$

$\tan\theta = \dfrac{10}{15} \Rightarrow \theta = 33.7^\circ$

$F_3 = \dfrac{15}{\cos\theta} = 18.0\text{N}$

Progress check

1 An object of mass 8 kg rests on a plane inclined at 40° to the horizontal. Find the components of its weight parallel and perpendicular to the plane. Take $g = 9.81 \text{ m s}^{-2}$.

2 Find the magnitude and direction of the resultant **R** N of the forces $(2\mathbf{i} + 3\mathbf{j})$ N, $(5\mathbf{i} - 4\mathbf{j})$ N and $(-3\mathbf{i} - 7\mathbf{j})$ N.

3 An object of weight 50 N rests on a rough horizontal surface. A horizontal force of 20 N is applied to the object so that it is on the point of slipping. Find the value of the coefficient of friction.

1 parallel: 50.4 N, perp: 60.1 N
2 8.94 N at 63.4° ↻ to **i**
3 0.4

3.4 Dynamics

After studying this section you should be able to:

- *understand and apply Newton's laws of motion in two or three dimensions*
- *analyse the motion of connected particles*
- *understand and apply the principle of conservation of momentum*
- *understand and use the coefficient of friction for moving particles*
- *use the relationship between impulse and momentum to solve problems*

LEARNING SUMMARY

Newton's laws of motion

Newton's laws of motion provide us with a clear set of rules that can be used to analyse the effect of forces within a system.

The laws may be stated as:

1. Every particle continues in a state of rest or uniform motion, in a straight line, unless acted upon by an external force.
2. The resultant force acting on a particle is equal to its rate of change of momentum. *This law is most often applied in the form* $F = ma$.
3. Every force has an equal and opposite reaction.

The formula F = ma

In the formula $F = ma$, F N stands for the resultant force acting on a particle, m kg is the mass of the particle and a m s^{-2} is the acceleration produced.

It is important to use the correct units.

Example

A particle of mass 2 kg rests on a smooth horizontal plane.
Horizontal forces of 15 N and 4 N act on the particle in opposite directions.
Find the acceleration of the particle.

Using $F = ma$

$$15 - 4 = 2a$$

$$\Rightarrow \quad a = 5.5$$

The acceleration of the particle is 5.5 m s^{-2}.

The formula applies equally well in two or three dimensions using vectors.

Example

Forces $(7\mathbf{i} - 2\mathbf{j} + \mathbf{k})$ N and $(3\mathbf{i} + \mathbf{j} - 5\mathbf{k})$ N act on a particle of mass 10 kg.
Find the acceleration produced.

The resultant force is the vector sum of the given forces, so using $\mathbf{F} = m\mathbf{a}$ gives

$$(7\mathbf{i} - 2\mathbf{j} + \mathbf{k}) + (3\mathbf{i} + \mathbf{j} - 5\mathbf{k}) = 10\mathbf{a}$$

$$\Rightarrow \quad 10\mathbf{i} - \mathbf{j} - 4\mathbf{k} = 10\mathbf{a}$$

$$\Rightarrow \quad \mathbf{a} = \mathbf{i} - 0.1\mathbf{j} - 0.4\mathbf{k}$$

The acceleration of the particle is $\mathbf{i} - 0.1\mathbf{j} - 0.4\mathbf{k}$ m s^{-2}.

The formula applies even when the force is variable.

Connected particles

In a typical problem, two particles are connected by a **light inextensible string**. Since the string is light, there is no need to consider its mass.
Since it is inextensible, both particles will have the same speed and accelerate at the same rate *while the string is taut*.

By Newton's third law, the tension in the string acts equally on both particles but in opposite directions.

To solve problems involving connected particles you need to:

- Draw a diagram.
- Identify and label the forces acting on each particle.
- Write the **equation of motion** for each particle i.e. apply $F = ma$ for each one.
- Solve the simultaneous equations produced.

Example
Two particles P and Q are connected by a light inextensible string. The particles are at rest on a smooth horizontal surface and the string is taut. A force of 10 N is applied to particle Q in the direction PQ. P has mass 2 kg and Q has mass 3 kg. Find the tension in the string and the acceleration of the system.

Note the use of the double headed arrow ⟶⟶ to represent acceleration.

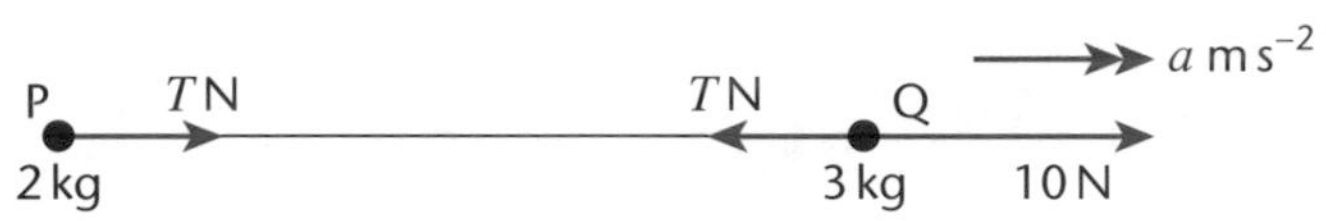

Using $F = ma$	For particle P	$T = 2a$	[1]
	For particle Q	$10 - T = 3a$	[2]
[1] + [2] gives		$10 = 5a \Rightarrow a = 2$	
Substituting for a in [1] gives		$T = 4$.	

The tension in the string is 4 N and the acceleration of the system is 2 m s^{-2}.

Momentum

The **momentum** of a particle is a vector quantity given by the product of its mass and its velocity i.e. momentum $= m\mathbf{v}$ where m kg is the mass of a particle and $\mathbf{v}$ m s^{-1} is its velocity.

The principle of conservation of momentum states that when two particles collide:

the total momentum before impact = the total momentum after impact.

This kind of diagram gives a useful way to represent the information.

Before impact

After impact

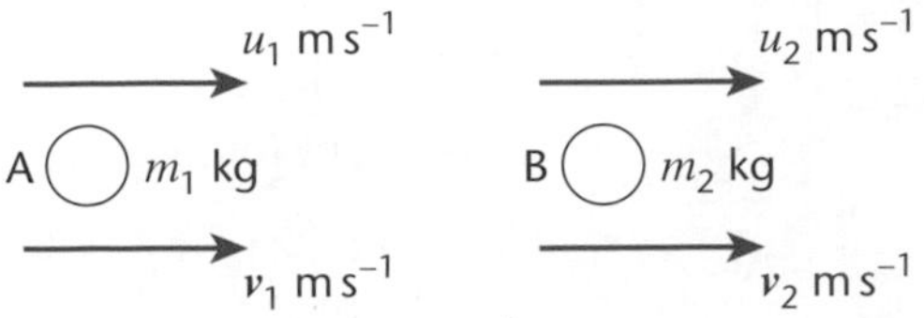

The positive direction for velocity is shown from left to right.

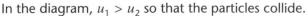

In the diagram, $u_1 > u_2$ so that the particles collide.

Conservation of momentum gives $m_1 u_1 + m_2 u_2 = m_1 v_1 + m_2 v_2$.

Example

A particle of mass 5 kg moving with speed 4 m s^{-1} hits a particle of mass 2 kg moving in the opposite direction with speed 3 m s^{-1}. After the impact the two particles move together with the same speed v m s^{-1}. Find the value of v.

Unknown velocities are shown in the positive direction.

Before impact

After impact

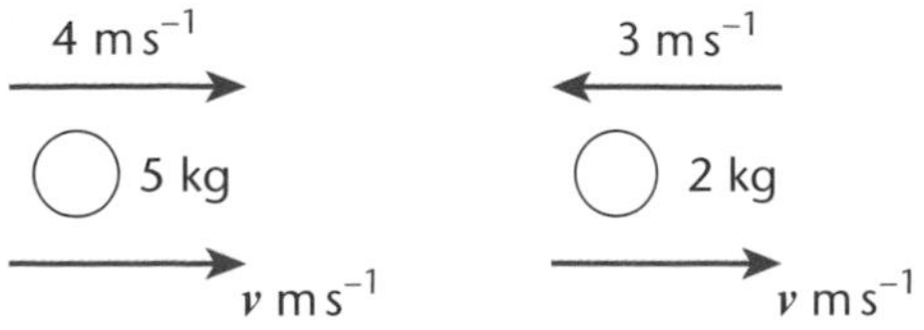

Conservation of momentum gives $5 \times 4 - 2 \times 3 = 5v + 2v$

$\Rightarrow \quad 7v = 14$

$\Rightarrow \quad v = 2.$

Impulse

The **impulse** of a constant force acting over a given time is given by the product of force and time. This may be written as $I = Ft$ where the impulse is I N s, the force is F N and the time is t s.

It follows that: Impulse = change in momentum

so $\quad Ft = mv - mu.$

The total momentum of the system is conserved.

When two particles collide, each receives an impulse from the other of equal size but opposite sign. In this way, the total change in momentum is zero as expected.

Example

A particle of mass 4 kg, initially at rest, is acted upon by a force of 3 N for 10 s. What is the speed of the particle at the end of this time?

Using $\quad Ft = mv - mu$

gives $\quad 3 \times 10 = 4v - 0$

$\Rightarrow \quad v = 7.5$

The speed of the particle is 7.5 m s^{-1}.

When two particles collide, the contact force between them may only last a very short time and is unlikely to be constant. However, the value of F in the equation $Ft = mv - mu$ may be used to represent the average value of this force.

Example

A ball of mass 0.5 kg strikes a hard floor with speed 2 m s^{-1} and rebounds with speed 1.5 m s^{-1}.

Given that the ball is in contact with the floor for 0.05 s find the average value of the contact force.

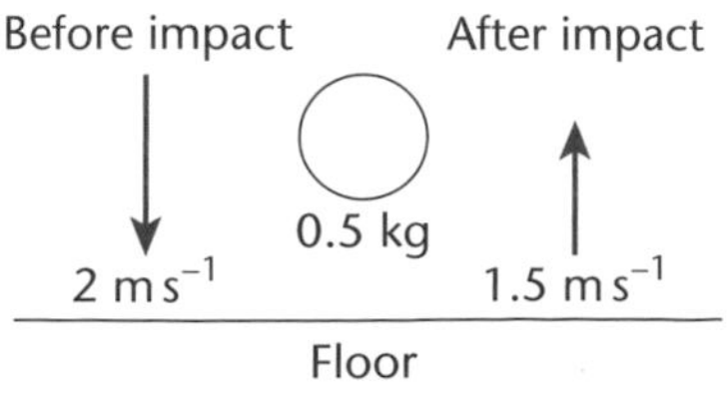

Taking upwards as the positive direction:

$u = -2$, $v = 1.5$, $m = 0.5$ and $t = 0.05$

Using $\quad Ft = mv - mu$

gives $\quad 0.05F = 0.5 \times 1.5 - 0.5 \times (-2)$

$\Rightarrow \quad 0.05F = 1.75$

$\Rightarrow \quad F = 35$

The average value of the contact force is 35 N.

Dynamic friction

When a particle moves over a rough surface, friction acts with its maximum value μR in the opposite direction to the motion.

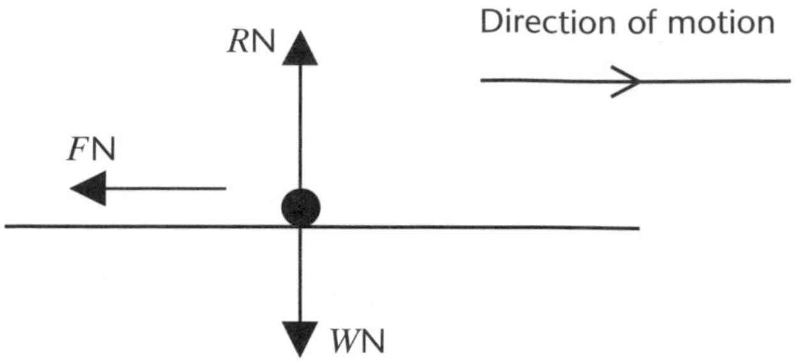

In the diagram , $F = \mu R$

Progress check

1 A particle of mass 8 kg is acted upon by the forces $(11\mathbf{i} - 3\mathbf{j})$ N, $(14\mathbf{i} + \mathbf{j})$ N and $(-\mathbf{i} + 10\mathbf{j})$ N. Find the acceleration of the particle.

2 Two particles A and B are connected by a light inextensible string. The particles are at rest on a smooth horizontal surface and the string is taut. A force of 12 N is applied to particle B in the direction AB. A has mass 4 kg and B has mass 2 kg.

Find the tension in the string and the acceleration of the system.

3 A particle of mass 8 kg moving with speed 2 m s^{-1} hits a stationary particle of mass 2 kg.
After the impact both particles move in the same direction with speed v m s^{-1}.

(a) Find the value of v.

(b) Find the impulse given to the stationary particle.

4 A particle of mass 0.2 kg moves over a rough horizontal surface with initial speed 3 ms^{-1}. The coefficient of friction between the particle and the surface is 0.25. How far does the particle travel before it is brought to rest?

1 $3\mathbf{i} + \mathbf{j}$ m s^{-2}.
2 Tension = 8 N, acceleration = 2 m s^{-2}.
3 (a) $v = 1.6$ (b) Impulse = 3.2 N s.
4 1.84 m (3 s.f.)

Sample questions and model answers

1

A snowboarder pushes off with an initial speed of 3 m s^{-1}, in a straight line. After 5 seconds, he is travelling at 5 m s^{-1}.

(a) Find an expression for his speed t seconds after the start.

(b) Find an expression for the distance travelled in t seconds.

(c) The length of the slope is 400 m. What is his speed at the bottom?

(a) $v = u + at \Rightarrow 5 = 3 + 5a \Rightarrow a = 0.4 \text{ m s}^{-2}$

$v = 3 + 0.4t$

(b) $s = ut + \frac{1}{2}at^2 \Rightarrow s = 3t + \frac{0.4}{2}t^2$

$s = 3t + 0.2t^2$

(c) $v^2 = u^2 + 2as \Rightarrow v^2 = 3^2 + 2 \times 0.4 \times 400$

$= 9 + 320$

$= 329$

$v = \sqrt{329}$

$= 18.2 \text{ m s}^{-1}$

Sample questions and model answers (continued)

2

Two particles are connected by a light inextensible string passing over a light frictionless pulley. Particle A has mass 5 kg and lies on a smooth horizontal table. Particle B has mass 4.5 kg and hangs freely. Take $g = 10 \text{ m s}^{-2}$.

(a) Write down the equation of motion for each particle.

(b) Find the acceleration of the system.

(c) Find the tension in the string.

A clearly labelled diagram is an essential first step.

Both particles have the same acceleration.

The tension is the same at both ends of the string.

(a)

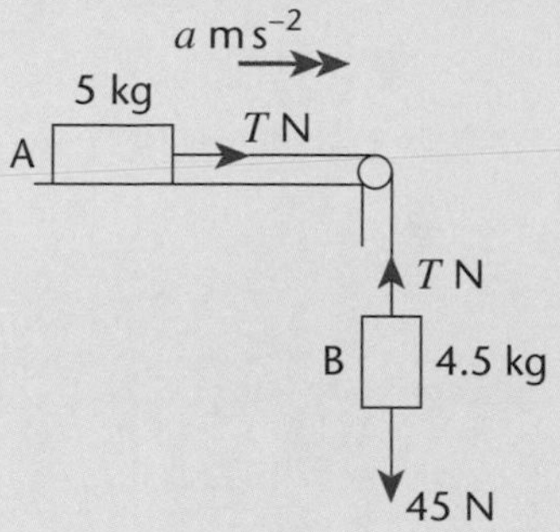

The diagram shows the forces acting on each particle in the direction of motion.

The equations of motion are:

Use $F = ma$ for each particle in turn.

for particle A $\quad T = 5a \quad$ [1]

for particle B $\quad 45 - T = 4.5a \quad$ [2]

Solve the simultaneous equations and interpret the results.

(b) [1] + [2] gives $\quad 45 = 9.5a$

$\Rightarrow a = 4.74$ to 3 s.f.

The acceleration of the system is 4.74 ms^{-2} to 3 s.f.

(c) Substitution for a in [1] gives T = 23.7 to 3 s.f.

It is a good idea to state the degree of accuracy given in your answer.

The tension in the string is 23.7 N to 3 s.f.

Practice examination questions

1 A particle moving in a straight line passes through a point O with velocity 10 m s^{-1} when t = 0. Given that the acceleration of the particle is –4 m s^{-2}.

Find:

(a) The velocity of the particle when $t = 5$.

(b) The distance that the particle travels between $t = 0$ and $t = 4$.

2

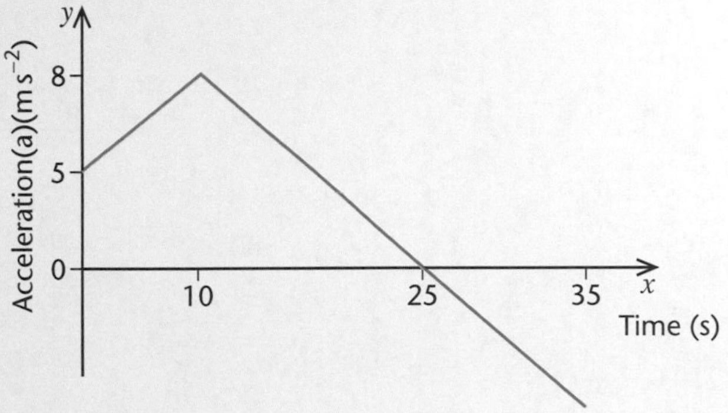

The diagram is an acceleration–time graph for the motion of a particle during a period of 35 s. The particle moves in a straight line and its initial velocity is 20 m s^{-1}.

(a) State the time at which the velocity reaches its maximum value.

(b) Calculate the maximum value of the velocity of the particle during this period.

3 A particle passes through a point P with position vector $(-5\mathbf{i} + 11\mathbf{j} + \mathbf{k})$ with velocity $(2\mathbf{i} + \mathbf{j})$ m s^{-1} when $t = 0$. The particle moves with constant acceleration $(\mathbf{i} + 3\mathbf{k})$ m s^{-2} and reaches the point Q when $t = 4$.

(a) Find $\overrightarrow{PQ}$.

(b) Find the position vector of Q.

(c) Find the average speed of the particle between P and Q.

4

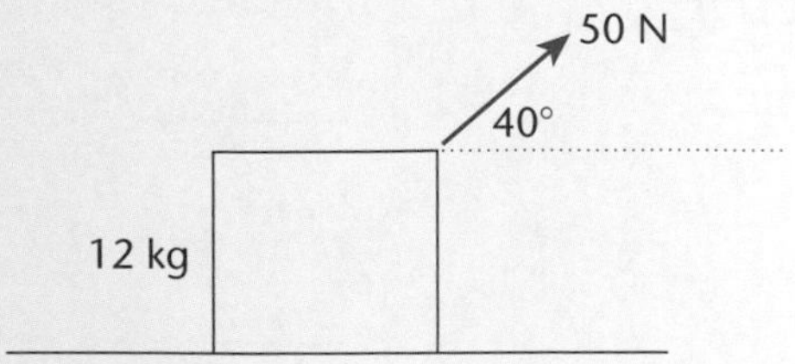

The diagram shows an object of mass 12 kg resting on a rough horizontal surface.

A force of 50 N acts on the object at an angle of 40^0 to the horizontal.
Take $g = 10$ m s^{-2}.

(a) Find the normal reaction between the horizontal surface and the object.

(b) Given that the object is in limiting equilibrium, find the coefficient of friction between the object and the horizontal surface.

Practice examination questions (continued)

5 The diagram shows a mass of 6 kg acted upon by a horizontal force of P N.

The coefficient of friction between the object and the plane is 0.3.

Find the minimum value of P required to maintain the object in equilibrium.

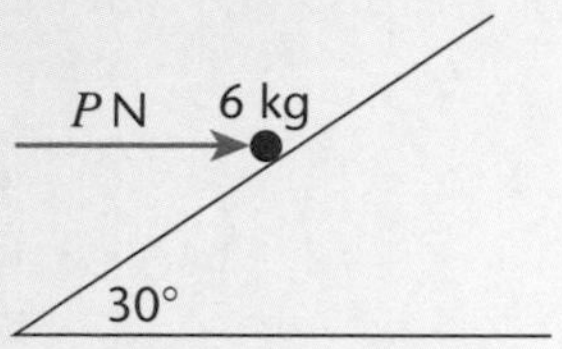

6

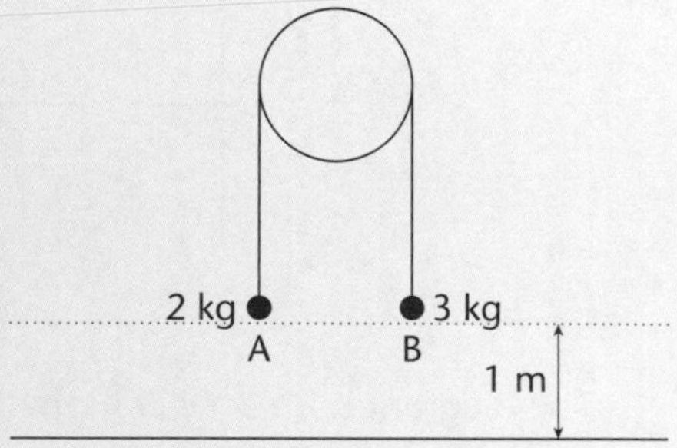

Particles A and B of mass 2 kg and 3 kg respectively are connected by a light inextensible string passing over a smooth frictionless pulley. The system is released from rest with both particles 1 m above the ground. Find:

(a) The acceleration of the system while the string remains taut.

(b) The tension in the string.

(c) The time taken for particle B to reach the ground.

(d) The speed with which particle B hits the ground.

(e) The greatest height reached by particle A.

(f) What difference would it make to your initial equations if the pulley was not light and frictionless?

7

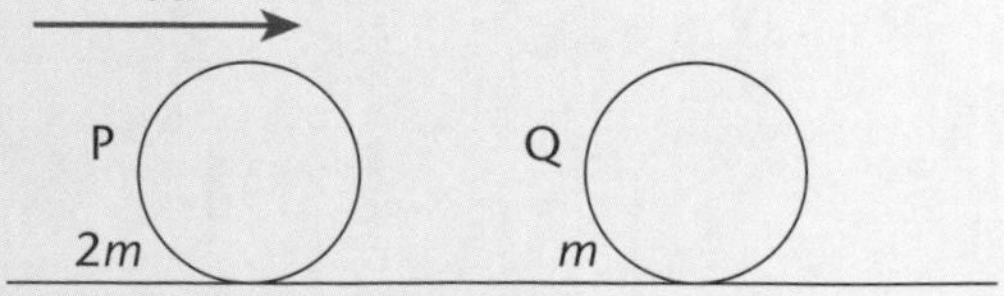

A particle P of mass $2m$ moving with speed $5u$ strikes particle Q of mass m, which is at rest.
After the collision both particles move in the same direction but the speed of particle Q is twice the speed of particle P.

(a) Find the speed of particle P after the collision.

(b) Find the magnitude of the impulse exerted on Q by P during the impact.

Chapter 4

Statistics 1

The following topics are covered in this chapter:

- *Representing data*
- *Probability*
- *Discrete random variables*
- *The normal distribution*
- *Correlation and regression*

4.1 Representing data

After studying this section you should be able to:

- *calculate averages for discrete and continuous data*
- *find the variance, standard deviation, range and inter-quartile range of data*
- *use diagrams to compare distributions*
- *interpret measures of location and dispersion in comparing sets of data*
- *understand the concepts of outliers and skewness*

LEARNING SUMMARY

Measures of central location

An average is a value that is taken to be representative of a data set. The three forms of average that you need are the mean, median and mode. These are often referred to as **measures of central location.**

The **mean** $\bar{x}$ of the values $x_1, x_2, x_3, ..., x_n$ is given by $\bar{x} = \frac{\Sigma x_i}{n}$. If each x_i occurs with frequency f_i then the mean of the **frequency distribution** is given by $\bar{x} = \frac{\Sigma f_i x_i}{\Sigma f_i}$.

Example
Find the mean of these results obtained by throwing a dice.

score	1	2	3	4	5	6
frequency	18	17	23	20	24	18

One problem with the mean is that it may be unduly influenced by a small number of extreme values, known as **outliers**.

$$\bar{x} = \frac{18 \times 1 + 17 \times 2 + 23 \times 3 + 20 \times 4 + 24 \times 5 + 18 \times 6}{18 + 17 + 23 + 20 + 24 + 18} = \frac{429}{120} = 3.575$$

The **median** is the middle value of the data when it is arranged in order of size. If there is an even number of values then the median is the mean of the middle pair. In the example above, the 60th and 61st values are both 4 so the median is 4.

The **mode** is the value that occurs with the highest frequency. In the example above the mode is 5.

You can *estimate* the mean of **grouped data** by using the midpoint of each class interval to represent the class.

Example
Estimate the mean value of h from the figures given in the table.
An estimate of the mean is given by

$$\bar{x} = \frac{755}{72} = 10.486...$$

10.5 = to 1 d.p.

Interval	frequency (f_i)	midpoint (x_i)	$f_i \times x_i$
$0 < h \leqslant 5$	8	2.5	20
$5 < h \leqslant 10$	24	7.5	180
$10 < h \leqslant 15$	29	12.5	362.5
$15 < h \leqslant 20$	11	17.5	192.5
Totals	72		755

Once data have been grouped, the exact values are not available and so it is only possible to *estimate* the mean.

Measures of dispersion

An average alone gives no indication of how widely dispersed the data values are. The simplest measure of dispersion, or spread, is the **range**.

The range of a data set is the difference between largest value and the smallest value.

A slightly more refined approach is to measure the spread of the 'middle half' of the data. The data is put into order and the values Q1, Q2 and Q3 are found that divide the data into quarters.

The diagram illustrates the situation for a data set of 11 values.

Q2 is the **median** of the data.

Q1, the **lower quartile**, is the median of the lower half of the data.

Q3, the **upper quartile**, is the median of the upper half of the data.

The **inter-quartile range** is then Q3 – Q1.

The quartiles may also be used to indicate whether the data values show **positive skew** (Q2 – Q1 $<$ Q3 – Q2) or **negative skew** (Q2 – Q1 $>$ Q3 – Q2).

The **variance** of a data set is the mean of the squares of the deviations from the mean.

The second form is easier to work with from raw data.

$$\text{Variance} = \frac{\Sigma (x_i - \bar{x})^2}{n} = \frac{\Sigma x_i^2}{n} - \bar{x}^2.$$

The variance gives an indication of the spread of data values about the mean but the units of the data have been squared in the process.

The value may be obtained directly from a scientific calculator but you may need to apply the formula.

$$\text{Standard deviation} = \sqrt{\text{variance}} = \sqrt{\frac{\Sigma x_i^2}{n} - \bar{x}^2}.$$

Standard deviation is measured in the same units as the data and is the value most commonly used to measure dispersion at this level.

For grouped data, take the mid-interval value for x.

In order to simplify calculations, data can be **coded**.

If a variable x is coded so that $y = \mathrm{a} + \mathrm{b}x$, then $\bar{y} = \mathrm{a} + \mathrm{b}\bar{x}$ and standard deviation $s_y = b \times s_x$

Statistical diagrams

The use of statistical diagrams may help in comparing distributions and revealing further information about the data.

A **box and whisker plot** for example shows the location and spread of a distribution at a glance.

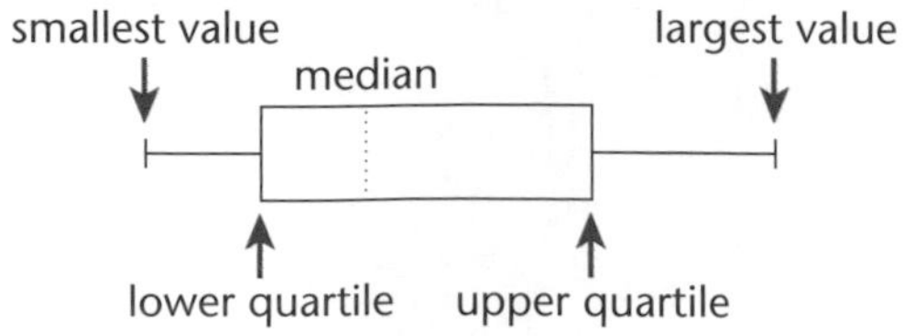

Stem and leaf diagrams have the advantage over bar charts that details of the data are available.

A **back-to-back stem and leaf diagram** allows direct comparison of two sets of data to be made.

The diagram gives a sense of location and spread for each data set.

The spread of marks is similar for the boys and the girls but the average for the girls is higher than for the boys.

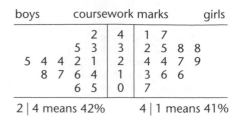

boys	coursework marks	girls
2	4	1 7
5 3	3	2 5 8 8
5 4 4 2 1	2	4 4 7 9
8 7 6 4	1	3 6 6
6 5	0	7

2 | 4 means 42% 4 | 1 means 41%

Histograms are useful for illustrating grouped continuous data.
The area of a bar is proportional to the frequency in that interval.
The height of the rectangle is given by the **frequency density**, where

$$\text{frequency density} = \frac{\text{frequency}}{\text{width of interval}}.$$

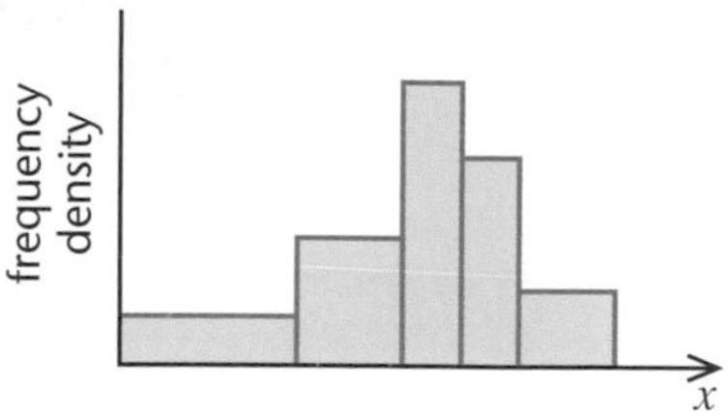

The **modal class** is the interval with the greatest frequency density. This is the highest bar on the histogram.

Joining with straight lines the mid points of the tops of the bars of a histogram gives a **frequency polygon**. Comparisons can be made by superimposing more than one frequency polygon on the same diagram.

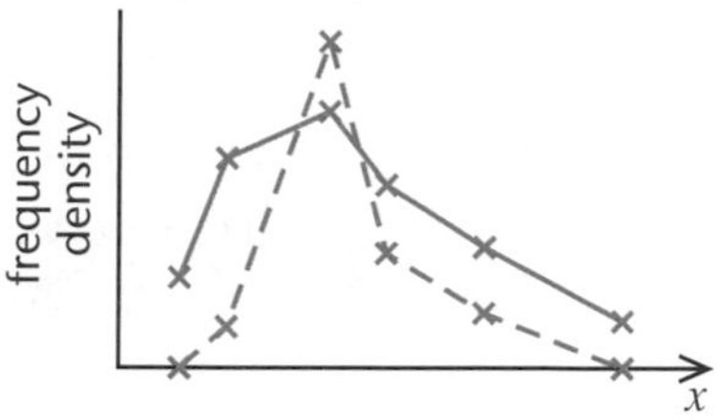

In a **cumulative frequency diagram**, cumulative frequencies (running totals) are plotted against upper class boundaries of the intervals. The median, Q2, and quartiles, Q1 and Q3, can be estimated from the graph. For n items,

Q1 is the value 25% ($\frac{1}{4}n$) through the data,
Q2 is the value 50% ($\frac{1}{2}n$) through the data,
Q3 is the value 75% ($\frac{3}{4}n$) through the data.

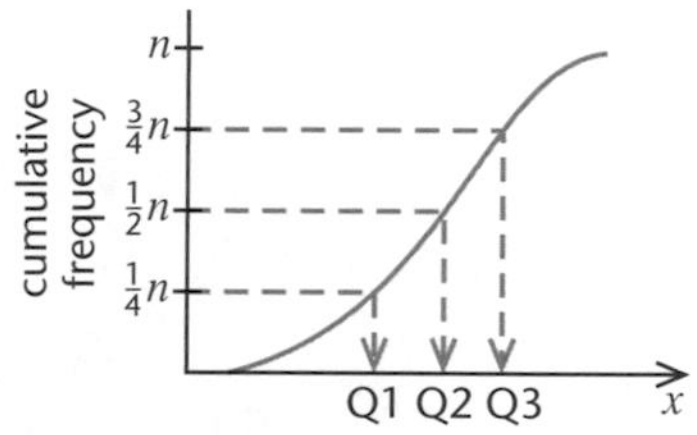

Progress check

1 For the numbers 8, 14, 20, 1, 6, 6, 13, 12, 4,
(i) find (a) the mean (b) the standard deviation (c) the median (d) the interquartile range,
(ii) draw a box and whisker plot.

2 Find the mean and standard deviation of this set of data:

x	0	1	2	3
f	4	6	7	3

3

Mass (g)	$20 \leqslant m < 40$	$40 \leqslant m < 50$	$50 \leqslant m < 55$	$55 \leqslant m < 60$	$60 \leqslant m < 90$
Frequency	10	12	10	7	9

For the above data
(a) estimate the mean and the standard deviation,
(b) draw a histogram and state the modal class,
(c) draw a cumulative frequency curve and use it to estimate
(i) the median (ii) the interquartile range.

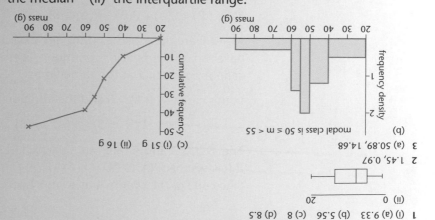

1 (i) (a) 9.33 (b) 5.56 (c) 8 (d) 8.5
(ii)

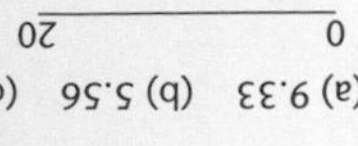

2 1.45, 0.97
3 (a) 50.89, 14.68
(b)

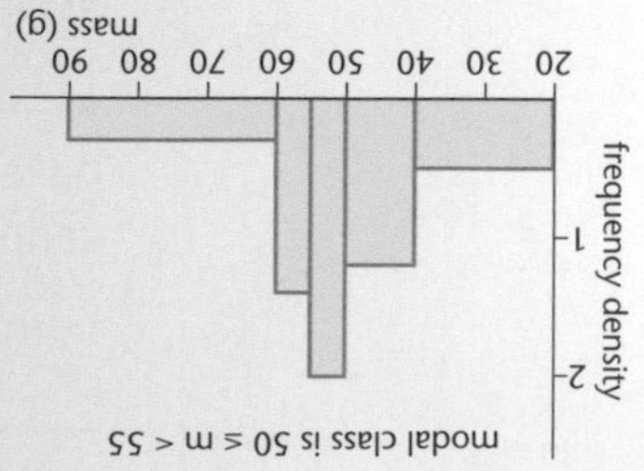

modal class is $50 \leqslant m < 55$

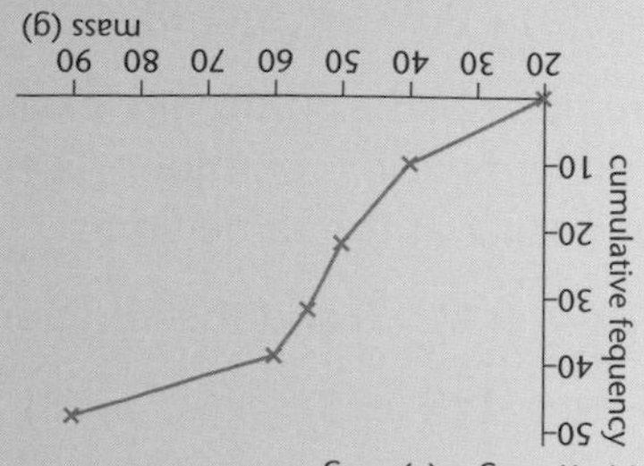

(c) (i) 51 g (ii) 16 g

4.2 Probability

After studying this section you should be able to:

- *use Venn diagrams to represent and interpret combined events*
- *use set notation to express and apply probability laws*
- *use tree diagrams to represent a problem*

LEARNING SUMMARY

Venn diagrams and probability laws

Venn diagrams provide a useful way to represent information about **sets** of objects.

The reason for mentioning them here is that the diagrams, and the notation used to express results, have a direct interpretation and application in probability theory.

In each diagram, the rectangle represents the set S of all objects under consideration.

A ∪ B is read as A union B.

The circles represent particular sets A and B of objects within the set S.

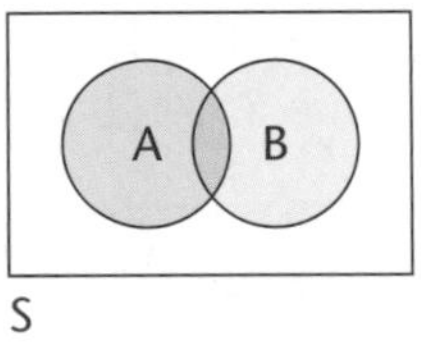

A ∪ B represents the set of all objects that belong to *either* A *or* B *or both.*

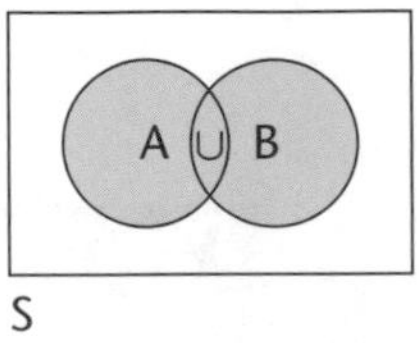

A ∩ B is read as A intersection B.

A ∩ B represents the set of those objects that belong to *both* A *and* B.

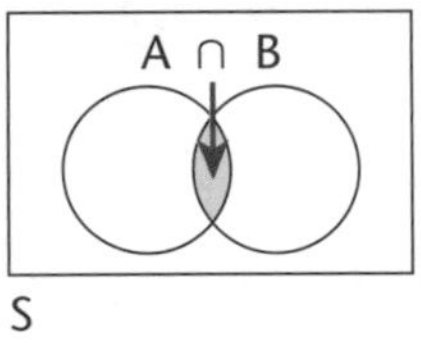

A′ represents the set of objects in S that do not belong to A.

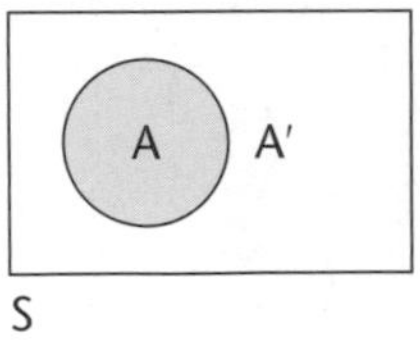

To make use of these ideas in probability theory:

- The objects are interpreted as **outcomes** for a particular situation.
- The set S is the **sample space** of all possible outcomes for the situation.
- The sets A and B are **events** defined by a particular choice of outcomes.
- P(A) for **example**, represents the probability that the event A occurs.
- P(A ∪ B) represents the probability that *either* A *or* B *or both* occurs.
- P(A ∩ B) represents the probability that *both* A *and* B occur.
- P(A′) represents the probability that A does *not* occur.

The **addition rule** states that

$P(A \cup B) = P(A) + P(B) - P(A \cap B)$.

A and B are described as **mutually exclusive events** if they have no outcomes in common. These are events that cannot both occur at the same time.

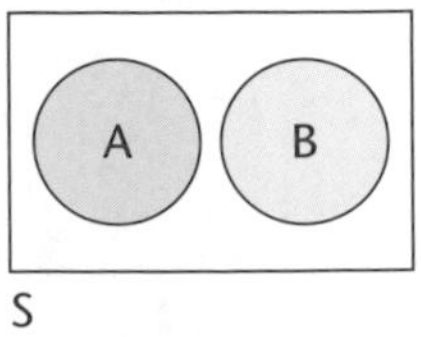

$P(A \cap B) = 0$

In this case $P(A \cap B) = 0$ and the addition law becomes

$$P(A \cup B) = P(A) + P(B).$$

The events A and A′ are mutually exclusive for any event A.
It follows that $P(A \cup A') = P(A) + P(A')$.
Since one of the events A or A′ must occur, $P(A \cup A') = 1$ so $P(A) + P(A') = 1$.

This is usually written as $P(A') = 1 - P(A)$.

The **multiplication rule** states that:

$$P(A \cap B) = P(A \mid B) \times P(B),$$

where $P(A \mid B)$ means the probability that A occurs *given that* B has *already* occurred.

$P(A \mid B)$ is described as a **conditional probability**, i.e. it represents the probability of A *conditional* upon B having occurred. Re-arranging the multiplication rule gives

$$P(A \mid B) = \frac{P(A \cap B)}{P(B)}.$$

If A and B are **independent events** then the probability that either event occurs is not affected by whether the other event has already occurred.

In this case, $P(A \mid B) = P(A)$ and the multiplication rule becomes

$$P(A \cap B) = P(A) \times P(B).$$

Example

The events A and B are independent. $P(A) = 0.3$ and $P(B) = 0.6$

Find (a) $P(A \cup B)$ (b) $P(A' \cap B)$.

(a) $P(A \cup B) = P(A) + P(B) - P(A \cap B)$.

Since A and B are independent, $P(A \cap B) = P(A) \times P(B)$.

So $P(A \cup B) = 0.3 + 0.6 - 0.3 \times 0.6 = 0.72$

(b) A and B are independent $\Rightarrow$ A′ and B are independent.

So $P(A' \cap B) = P(A') \times P(B) = (1 - 0.3) \times 0.6 = 0.42$

In some problems you need to define the events first before you can apply the probability rules.

A **tree diagram** is a useful way to represent the probabilities of combined events. Each path through the diagram corresponds to a particular sequence of events and the multiplication rule is used to find its probability. When more than one path satisfies the conditions of a problem, these probabilities are added.

This is true whether or not the events are independent.

Example

The probability that Chris gets grade A for Maths is 0.6 and the corresponding probability for English is 0.7. The events are independent.

(a) Calculate the probability that Chris gets just one A in the two subjects.
(b) Given that Chris gets just one grade A, find the probability that it is for Maths.

Show the information on a tree diagram.

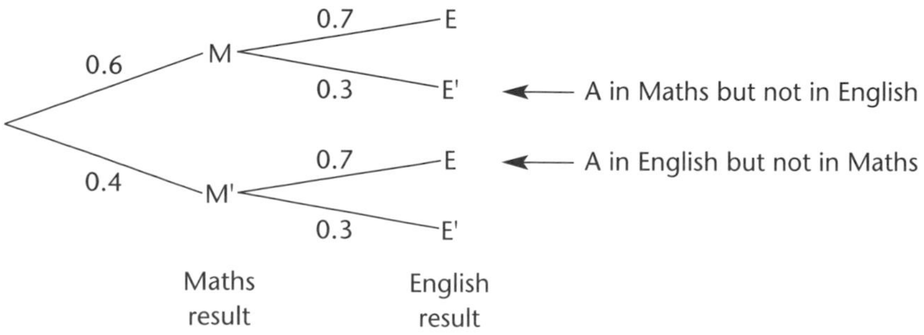

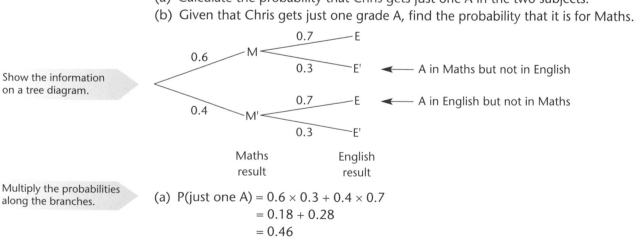

Multiply the probabilities along the branches.

(a) $P(\text{just one A}) = 0.6 \times 0.3 + 0.4 \times 0.7$
$= 0.18 + 0.28$
$= 0.46$

Use conditional probability.

(b) $P(\text{A in Maths} \mid \text{just one A}) = \dfrac{P(\text{A in Maths and just one A})}{P(\text{just one A})}$

$= \dfrac{0.18}{0.46}$

$= 0.39$ (2 d.p.)

Progress check

1 The probability that the event B occurs is 0.7. The probability that events A and B both occur is 0.4. What is the probability that A occurs given that B has already occurred?

2 It is given that $P(A) = 0.4$, $P(B) = 0.7$ and $P(A \cap B) = 0.2$
Find (a) $P(A \cup B)$ (b) $P(B \mid A)$ (c) $P(A \mid B)$.

3 A bag contains 3 red counters and 7 green counters. Two counters are taken at random from the bag, without replacement.
(a) By drawing a tree diagram, or otherwise, find the probability that the counters are different colours.
(b) Given that the counters are different colours, find the probability that the first counter picked is red.

1 4/7
2 (a) 0.9 (b) 0.5 (c) $\frac{2}{7}$
3 (a) $\frac{7}{15}$ (b) $\frac{1}{2}$

4.3 Discrete random variables

After studying this section you should be able to:

- *understand the concept of a discrete random variable, its mean and variance*
- *calculate the expectation and variance of functions of a random variable*
- *understand the discrete uniform distribution*

LEARNING SUMMARY

Discrete random variables

The score obtained when a dice is thrown may be thought of as a **random variable**. Since only specific values may be obtained, this is an example of a **discrete random variable**. In the usual notation:

- A capital letter such as X is used as a label for the random variable.
- A lower case letter such as x is used to represent a particular value of X.
- The probability that X takes the value x is written as $P(X=x)$ or just $p(x)$ and this is known as the **probability function**.
- If X is a discrete random variable with probability function $p(x)$ then $\sum p(x) = 1$.

This is a useful result.

The **probability distribution** of X is given by the set of all possible values of x together with the values of $p(x)$. This is usually shown in a table.

For example, the probability distribution of the scores shown on a dice may be given as:

This is an example of a discrete *uniform* distribution, as all the probabilities are the same.

x:	1	2	3	4	5	6
$p(x)$:	$\frac{1}{6}$	$\frac{1}{6}$	$\frac{1}{6}$	$\frac{1}{6}$	$\frac{1}{6}$	$\frac{1}{6}$

The **cumulative distribution function** is given by $F(x_0) = P(X \leqslant x_0) = \sum_{x \leqslant x_0} p(x)$.

This represents the probability that the random variable X takes a value less than or equal to x_o and is given by the sum of the probabilities up to and including that point.

Mean and variance

The symbol μ is used to stand for the **mean value of X**. This is also known as the **expected value, or expectation,** of X and is written $E(X)$.
For any discrete random variable X the mean is $\mu = E(X) = \sum xp(x)$.

$E(X) = \sum xP(X = x)$

Example Find the expected value of the score shown on a dice.

x:	1	2	3	4	5	6
$p(x)$:	$\frac{1}{6}$	$\frac{1}{6}$	$\frac{1}{6}$	$\frac{1}{6}$	$\frac{1}{6}$	$\frac{1}{6}$
$xp(x)$:	$\frac{1}{6}$	$\frac{2}{6}$	$\frac{3}{6}$	$\frac{4}{6}$	$\frac{5}{6}$	$\frac{6}{6}$

$$\mu = \tfrac{1}{6} + \tfrac{2}{6} + \tfrac{3}{6} + \tfrac{4}{6} + \tfrac{5}{6} + \tfrac{6}{6} = \tfrac{21}{6}$$

The expected value is 3.5

$E(g(X)) = \Sigma g(x)\, P(X = x)$

The **expected value of a function of a random variable** is found in a similar way. In general, if $g(X)$ is some function of the random variable X then

$$E(g(X)) = \Sigma g(x)p(x).$$

$E(X^2) = \Sigma x^2 P(X = x)$

For example, $E(X^2) = \Sigma\, x^2 p(x)$. In the case of the dice scores above this gives:

$$E(X^2) = \tfrac{1}{6} + \tfrac{4}{6} + \tfrac{9}{6} + \tfrac{16}{6} + \tfrac{25}{6} + \tfrac{36}{6} = \tfrac{91}{6}\ .$$

Using σ to stand for the **standard deviation of** X and $\mathrm{Var}(X)$ to stand for the **variance of** X we have:

$$\sigma^2 = \mathrm{Var}(X) = E(X^2) - \mu^2.$$

Referring to the example of the dice scores again:
$\mathrm{Var}(X) = \tfrac{91}{6} - (\tfrac{21}{6})^2 = \tfrac{35}{12}$ and $\sigma = \sqrt{\tfrac{35}{12}}$.

The **expectation of a linear function of X** can be expressed in terms of $E(X)$.

In general: $E(aX + b) = aE(X) + b$.

Once the value of $E(X)$ is known, applying the result above is much simpler than working out $\Sigma\,(ax + b)p(x)$. However, the result only holds for linear functions, so for example $E(X^2)$ is not the same as $(E(X))^2$.

The **variance of a linear function of** X may be expressed in terms of $\mathrm{Var}(X)$.

In general: $\mathrm{Var}(aX + b) = a^2\, \mathrm{Var}(X)$.

Progress check

1 The random variable X has the probability distribution shown. Find the value of k.

x:	0	1	2
$p(x)$:	0.3	0.2	k

2 Find (a) $E(X)$ (b) $\mathrm{Var}(X)$ for the probability distribution in question 1.

3 The random variable Y is given by $Y = 3X + 2$, where X is as in question 1.
Find (a) $E(Y)$ (b) $\mathrm{Var}(Y)$.

4 X has a discrete uniform distribution where $P(X = x) = k$ for $x = 30, 40, 50, 60, 70$.
Find the value of k and the mean and standard deviation of X.

1 $k = 0.5$
2 (a) 1.2 (b) 0.76
3 (a) 5.6 (b) 6.84
4 0.2, 50, 14.1

4.4 The normal distribution

After studying this section you should be able to:

- *use the normal distribution model to calculate probabilities*

LEARNING SUMMARY

The normal distribution

The **normal variable** is a continuous random variable. Its probability function is represented by a bell-shaped curve, symmetric about the mean value μ.

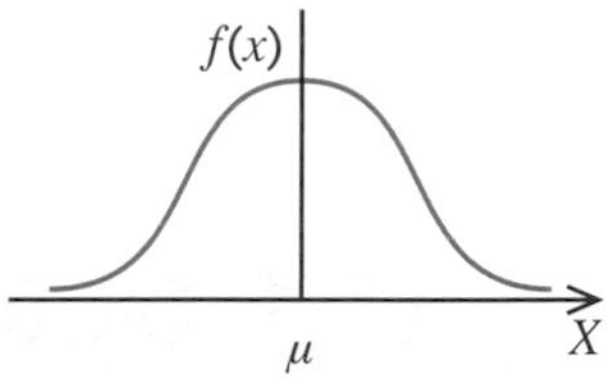

The parameters of the normal distribution are μ and σ^2. If the continuous random variable X follows a normal distribution with these **parameters** then this is written as $X \sim N(\mu, \sigma^2)$. The standard deviation is σ.

Tabulated values of probabilities for the normal distribution use Z as the variable.

The **standard normal variable** ***Z*** is used to calculate probabilities based on the Normal distribution where $Z = \dfrac{X - \mu}{\sigma}$ and $Z \sim \text{N}(0, 1)$.

Example

Find $\text{P}(15 < X < 17)$ where $X \sim \text{N}(15, 25)$

$$15 < X < 17 \Rightarrow \frac{15 - 15}{5} < Z < \frac{17 - 15}{5} \Rightarrow 0 < Z < 0.4$$

Using the normal distribution tables.

$$\begin{aligned} \text{P}(15 < X < 17) &= \text{P}(0 < Z < 0.4) \\ &= \Phi(0.4) - \Phi(0) = 0.6554 - 0.5000 \\ &= 0.1554 \end{aligned}$$

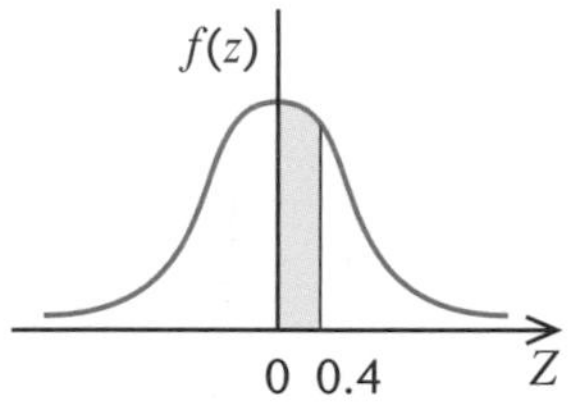

You may need to use given information to set up and solve simultaneous equations to find the values of μ and σ.

Example

It is given that $\text{P}(X > 70) = 0.0436$ and $\text{P}(X < 45) = 0.3669$, where $X \sim N(\mu, \sigma^2)$. Find the values of μ and σ.

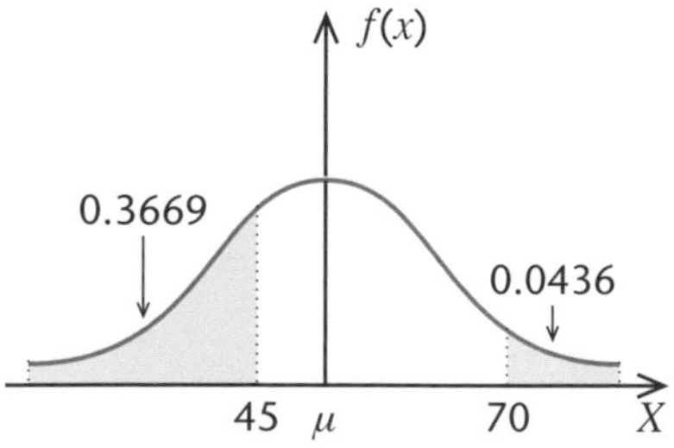

$$P(X > 70) = P\left(Z > \frac{70 - \mu}{\sigma}\right) = 0.0436$$

$$\Phi\left(\frac{70 - \mu}{\sigma}\right) = 1 - 0.0436 = 0.9564$$

From the tables

$$\left(\frac{70 - \mu}{\sigma}\right) = 1.71 \Rightarrow 70 - \mu = 1.71\sigma \quad [1]$$

Similarly

$$P(X < 45) = P\left(Z < \frac{45 - \mu}{\sigma}\right) = 0.3669$$

$$\Phi\left(\frac{45 - \mu}{\sigma}\right) = 0.3669$$

In general, $\Phi(-z) = 1 - \Phi(z)$.

$$\Phi\left(-\left(\frac{45 - \mu}{\sigma}\right)\right) = 1 - 0.3669 = 0.6331$$

$$-\left(\frac{45 - \mu}{\sigma}\right) = 0.34$$

$$45 - \mu = -0.34\sigma \qquad [2]$$

[1] – [2] gives $25 = 2.05\sigma \Rightarrow \sigma = 12.2$

Substituting in (1): $\mu = 49.1$

Progress check

1 Find $P(17 < X < 24)$ where $X \sim N(14, 16)$.

2 It is given that $X \sim N(\mu, \sigma^2)$ and $P(X < 30) = 0.7$.
Given also that $P(X > 35) = 0.1$, find μ and σ.

1 0.2204
2 $\mu = 26.5$
$\sigma = 6.60$

4.5 Correlation and regression

After studying this section you should be able to:

- *calculate and interpret the product-moment correlation coefficient*
- *recognise explanatory and response variables*
- *find the equations of least squares regression lines*
- *use a suitable regression line to estimate a value*

LEARNING SUMMARY

All of the work in this section relates to the treatment of **bivariate data**. This is data in which each data point is defined by two variables.

Correlation

A **scatter diagram** may be used to represent bivariate data. The extent to which the points approximate to a straight line gives an indication of the strength of a linear relationship between the variables, known as the **linear correlation.**

You should only use this method when both variables are normally distributed.

One way to arrive at a numerical measure of the correlation is to use the **product–moment correlation coefficient,** r.

For n pairs of (x, y) values:

$$S_{xx} = \Sigma(x - \bar{x})^2 = \Sigma x^2 - \frac{(\Sigma x)^2}{n} \qquad S_{yy} = \Sigma(y - \bar{y})^2 = \Sigma y^2 - \frac{(\Sigma y)^2}{n}$$

$$S_{xy} = \Sigma(x - \bar{x})(y - \bar{y}) = \Sigma xy - \frac{(\Sigma x)(\Sigma y)}{n}$$

and the product–moment correlation coefficient is given by:

You may be able to obtain the value of r directly from your calculator.

$$r = \frac{S_{xy}}{\sqrt{S_{xx}S_{yy}}}$$

This gives values of r between –1 (representing a perfect negative correlation) and +1 (representing a perfect positive correlation).

Regression

Whereas correlation is determined by the *strength* of a linear relationship between the two variables, **regression** is about the *form* of the relationship given by the equation of a **regression line.** (Make sure you known the difference between correlation and regression.)

The purpose in establishing the equation of a regression line is to make predictions about the values of one variable (known as the **response variable**) for some given values of the other variable (known as the **explanatory variable**).

Predictions should only be made for values within the range of readings of the explanatory variable. **Extrapolation** for values outside this range is unreliable. Another factor affecting the accuracy of any predictions is the influence of **outliers** on the equation of the regression line.

Figure 1 illustrates **y-residuals** given by

d = (observed value of y) – (predicted value of y).

Figure 2 illustrates **x-residuals** given by

d = (observed value of x) – (predicted value of x).

A **least squares regression line** is a line for which the sum of the squares of either the x-residuals or y-residuals is minimised.

Both lines always pass through the point $(\bar{x}, \bar{y})$.

Figure 1

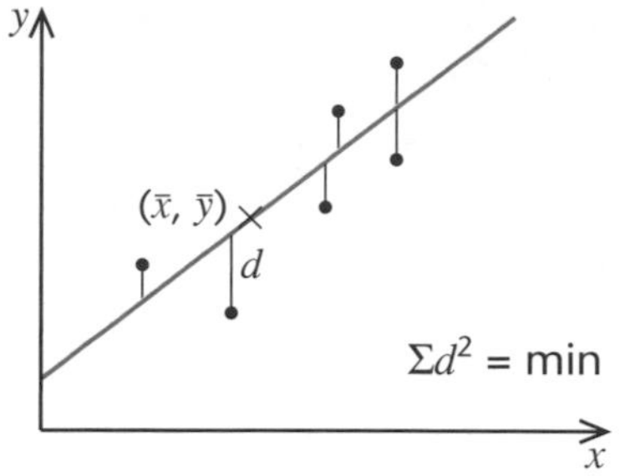

Figure 2

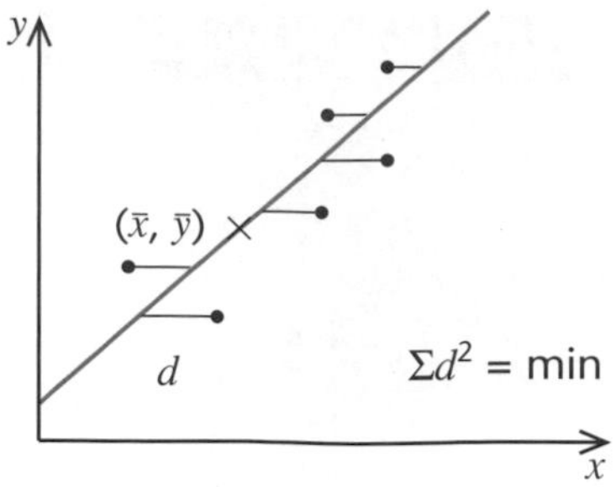

This gives the regression line of y on x as

$$y = a + bx,$$

where $b = \dfrac{S_{xy}}{S_{xx}}$ and $a = \bar{y} - b\bar{x}$.

Use this equation to estimate values of y for given values of x when x is the explanatory variable and y is the response variable.

This gives the regression line of x on y as

$$x = c + dy,$$

where $d = \dfrac{S_{xy}}{S_{yy}}$ and $c = \bar{x} - d\bar{y}$.

Use this equation to estimate values of x for given values of y when y is the explanatory variable and x is the response variable.

The values of a and b may be obtained directly from some calculators.

Unless there is perfect correlation between the variables, the two regression lines will be different and you cannot rearrange one equation to obtain the other.

Progress check

The summary data for 10 pairs of (x, y) values is as follows:

$$\sum x = 146, \quad \sum x^2 = 2208, \quad \sum y = 147, \quad \sum y^2 = 2247, \quad \sum xy = 2211.$$

1 Find the value of the product–moment correlation coefficient between x and y.
2 Find the equation of the least squares regression line of y on x.

1 0.799
2 $y = 2.317 + 0.848x$

Sample questions and model answers

Scientific calculators can find the standard deviation from the raw data, but you need to know the formulae because the data may be given in summarised form.

1

Paul swims 20 lengths of a swimming pool in training for a competition. His times, x seconds, for completing lengths of the pool are summarised by:

$$\sum x = 372 \qquad \sum x^2 = 6952.$$

(a) Find the mean and standard deviation of Paul's times for completing one length of the pool.

(b) A month ago, Paul's mean time per length was 19.8 s with a standard deviation of 1.72 s. Comment on the change in his performance indicated by these results.

(a)
$$\bar{x} = \frac{\sum x}{n} = \frac{372}{20}$$
$$= 18.6$$

Paul's mean time per length is 18.6 s

Quote the formula and substitute the information.

$$\sqrt{\frac{\sum x^2}{n} - \bar{x}^2} = \sqrt{\frac{6952}{20} - 18.6^2}$$
$$= 1.28$$

The standard deviation of Paul's times is 1.28 s

(b) The mean time has been reduced, suggesting an improvement in performance. The standard deviation has also been reduced which suggests a greater consistency of performance.

You may be expected to interpret results within the context of the question and make comparisons.

2

A machine produces nails with lengths that are normally distributed with mean 5.2 cm and standard deviation 0.06 cm.

Find the probability that a nail selected at random has a length between 5.1 cm and 5.25 cm.

Define the random variable X.

Let X be the length, in cm, of a nail, where $X \sim N(5.2, 0.06^2)$

Express the probability in terms of X and then standardise.

$$P(5.1 < X < 5.25) = P\left(\frac{5.1 - 5.2}{0.06} < Z < \frac{5.25 - 5.2}{0.06}\right)$$
$$= P(-1.67 < Z < 0.83)$$
$$= \Phi(0.83) - \Phi(-1.67)$$
$$= \Phi(0.83) - (1 - \Phi(1.67))$$
$$= 0.7967 - (1 - 0.9525)$$
$$= 0.7492$$

It follows from the symmetry of the distribution that $\Phi(-z) = 1 - \Phi(z)$.

The probability that a nail selected at random has a length between 5.1 cm and 5.25 cm is 0.749 (3 s.f.)

Practice examination questions

1 A discrete random variable X has the probability function p(x) given by

$$p(x) = \frac{x^2}{k} \qquad \text{for} \quad x = 1, 2, 3.$$

$$p(x) = 0 \qquad \text{otherwise.}$$

(a) Find the value of k.

(b) Calculate the mean and variance of the probability distribution.

(c) Find the mean and variance of the random variable $Y = 3X - 5$.

2 The events A and B are such that

$$P(A) = 0.3 \qquad P(B) = 0.6 \qquad P(A \cup B) = 0.72$$

(a) Show that the events A and B are not mutually exclusive.

(b) Show that the events A and B are independent.

(c) Write down $P(A \cup B)'$.

(d) Find $P(A' \cup B')$.

3 (a) Three cards are drawn at random, without replacement, from a pack of playing cards.
Calculate the probability that:

(i) All three cards are red.

(ii) All three cards are the same colour.

(iii) At least one of the cards is red.

(b) The cards are returned to the pack and the pack is shuffled. One more card is drawn.

Find the probability that this card is an ace, given that it is not a diamond.

4 The lifetime of batteries used in a torch can be modelled by a Normal distribution with a mean of 8 h and a standard deviation of 25 mins.

(a) Calculate the probability that a battery selected at random will last for more than $8\frac{1}{2}$ h.

(b) Find the percentage of batteries that will last for less than 7 h.

Practice examination questions (continued)

5 Eleven students in the sixth form of a school study both mathematics and physics at A Level. It is thought that their physics scores depend on their maths scores. The scores in an end of term test are shown for 10 of these students.

Maths (x)	48	39	52	55	71	62	76	54	61	58
Physics (y)	41	35	50	47	63	48	70	45	46	39

Using x to represent the score in maths and y to represent the corresponding score in physics:

$$\Sigma x = 576,\ \Sigma y = 484,\ \Sigma x^2 = 34216,\ \Sigma y^2 = 24450,\ \Sigma xy = 28785$$

(a) Calculate the product–moment correlation coefficient, between x and y.

(b) Find the equation of the regression line of y on x for the data.

(c) The eleventh student obtained 65 on the maths test. Estimate this student's score in the physics test, commenting on the reliability of your estimate.

Chapter 5

Decision Mathematics 1

The following topics are covered in this chapter:

- *Algorithms*
- *Graphs and networks*
- *Critical path analysis*
- *Linear programming*
- *Matchings*

5.1 Algorithms

After studying this section you should be able to:

LEARNING SUMMARY

- *implement an algorithm given by text or flow chart*
- *implement the following algorithms from memory:*
  - *bubble sort*
  - *quick sort*
  - *binary search*
  - *bin packing*

Introducing algorithms

An **algorithm** is a finite sequence of precise instructions used to solve a problem.

One way to define an algorithm is to simply list all of the necessary steps. In more complex situations, the process may be easier to follow if the algorithm is presented as a **flow diagram**.

Algorithms play a fundamental role throughout Decision Mathematics.

You need to be able to implement an algorithm presented in either form.

In the exam you may be asked to implement an unfamiliar algorithm and interpret the results.

Example

Implement the algorithm given by the flow chart and state what the written values represent.

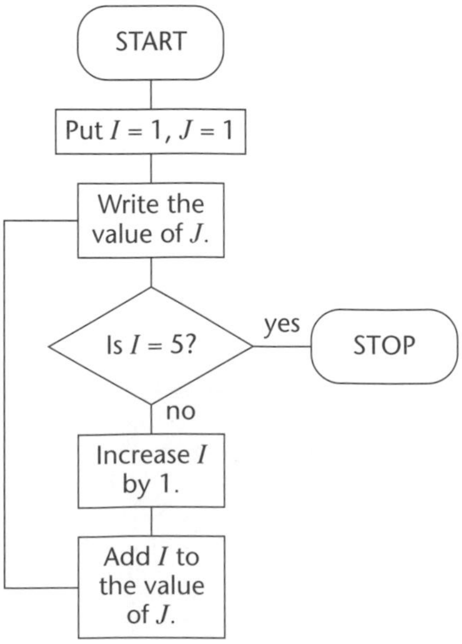

The table shows how the values of *I* and *J* change as the algorithm is implemented.

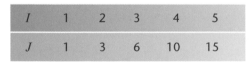

I	1	2	3	4	5
J	1	3	6	10	15

The written values are 1, 3, 6, 10, 15 and these are the first five triangle numbers.

The algorithm may be expressed in words as:

Write down a sequence of 5 numbers starting with 1 and then increasing, first by 2 and then by 1 more each time from term to term.

This version is more concise but not so easy to understand.

In the exam you may be expected to implement any of these algorithms without being reminded of the necessary steps.

You need to know and be able to implement the following algorithms.

The bubble sort

As its name suggests, the **bubble sort** is an algorithm for sorting a list in a particular order.

Step 1 Compare the first two elements of the list and switch them if they are in the wrong order.

Step 2 Compare the second and third elements of the list and, again, switch them if they are in the wrong order.

Step 3 Continue in the same way until you reach the end of the list. This completes the first **pass** through the list.

Step 4 Make repeated passes through the list until a pass produces no change.

Example Sort the list 5, 7, 3, 11, 6, 8 into ascending order.

The first pass gives

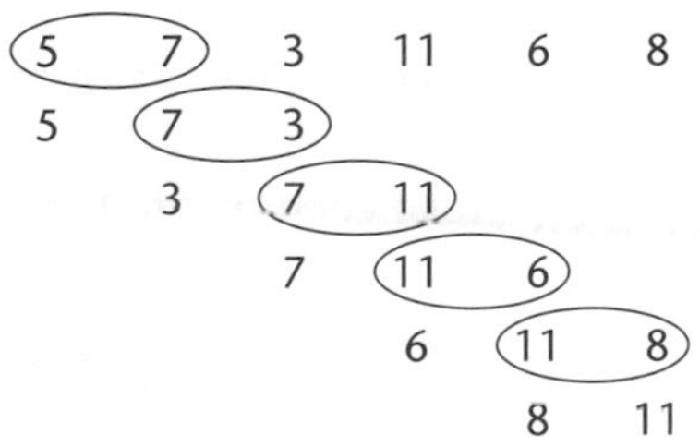

The second pass produces 3, 5, 6, 7, 8, 11.

The third pass produces 3, 5, 6, 7, 8, 11.

Don't just look at the list and say that the numbers are now in order. Use the algorithm to decide when to stop.

The third pass made no change so the ordering is complete.

The quick sort

This algorithm relies on finding the middle element of a list. For n elements the middle element has position $\frac{1}{2}(n+1)$ when n is odd and $\frac{1}{2}(n+2)$ when n is even.

Step 1 Take the middle element as the **pivot**.

Step 2 Consider each of the other elements in turn and place any with order less than or equal to the pivot on its one side and any greater than the pivot on its other side. Do this without re-ordering the numbers on either side of the pivot.

This creates two sub-lists with the pivot in-between.

Step 3 Repeat steps 1 and 2 for each sub-list until each one contains a single element.

Example Sort the list 4, 2, 6, 10, 5, 7 into ascending order.

4, 2, 6, 5, 7, 10

4, 2, 5, 6, 7, 10

2, 4, 5, 6, 7, 10

2, 4, 5, 6, 7, 10

Values shown highlighted have been used as pivots.

It is easy to see that the list is now in order but the sub-list 4, 5 has more than one element, so you must continue with the algorithm.

The ordered list is 2, 4, 5, 6, 7, 10.

Binary search

The binary search algorithm provides a systematic way of searching through an ordered list to find an element that matches your criterion. **For example**, you may need to look through a set of index cards to find the contact details of a customer.

This is not a very efficient method if the required element is at the end of the list.

Step 1 Find the element in the middle of the list. If this element matches your criterion then stop. If it does not, then use the middle term to decide in which half of the list the required element must lie.

Step 2 Repeat Step 1 for the selected half.

Bin packing

The bin packing algorithm is used to solve problems that can be represented by the need to pack some boxes of equal cross-section but different heights into bins, with the same cross-section as the boxes, using as few bins as possible.

There is no known algorithm that will always provide the *best* solution. You need to be familiar with the three algorithms below that attempt to provide a *good* solution. Such algorithms are known as **heuristic** algorithms.

First-fit algorithm

Take each box in turn from the order given and pack it into the first available bin.

Full-bin algorithm

The full-bin algorithm is only practical when the number of bins and boxes is small.

Step 1 Considering the boxes in the given order, use the first available combination that will fill a bin.

Step 2 Repeat Step 1 until no more bins can be filled.

Step 3 Implement the first-fit algorithm for the remaining boxes.

First-fit decreasing algorithm

Step 1 Arrange the boxes in decreasing order of size.

Step 2 Implement the first-fit algorithm starting with the largest box.

Progress check

1 Write the letters P, Q, B, C, A, D in alphabetical order using:

(a) the bubble sort algorithm

(b) the quick sort algorithm.

2 A project involves activities A–H with durations in hours as given in the table.

A	B	C	D	E	F	G	H
3	1	5	4	2	3	4	2

You can represent each worker as a 'bin' of 'height' 8 hours.

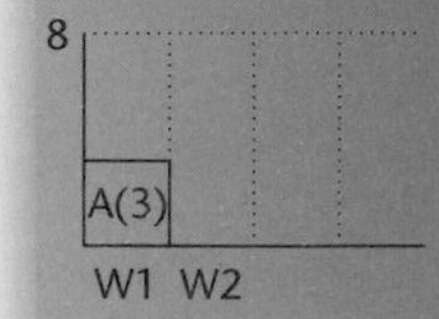

The project is to be completed in 8 hours.

(a) Use the first-fit algorithm to try to find the minimum number of workers needed.

(b) Find a better solution.

1 A, B, C, D, P, Q.

2 First-fit: 4 workers; A better solution is (A, B, D) (C, F) (E, G, H).

5.2 Graphs and networks

After studying this section you should be able to:

- *understand the use of terminology associated with graphs and networks*
- *use Prim's and Kruskal's algorithms for constructing a minimum spanning tree*
- *use Dijkstra's algorithm to find a shortest path between two vertices*
- *use the route inspection algorithm to find a route of minimum weight that traverses every arc of a network*

LEARNING SUMMARY

Defining terms

- A **graph** is a set of points, called **vertices** or **nodes**, connected by lines called **edges** or **arcs**.
- A **simple graph** is one that has no loops and in which no pair of vertices are connected by more than one edge.

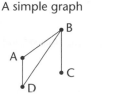

It is worth spending time getting to know all of the terms. Their definitions are quite detailed so you will need to keep reminding yourself of what each term means.

A simple graph

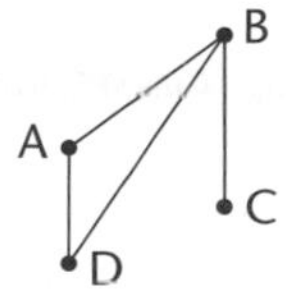

A non-simple graph

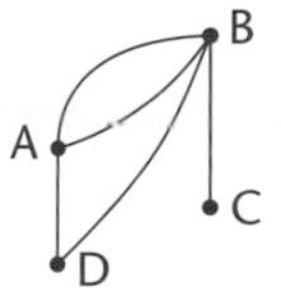

Two edges between A and B

A non-simple graph

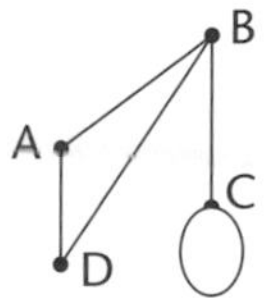

A loop around C

- The number of edges incident on a vertex is called its **order**, **degree** or **valency**. A vertex may be **odd** or **even** depending on whether its order is odd or even.
- A **subgraph** of some graph G is a graph consisting entirely of vertices and edges that belong to G.
- A **directed** edge is an edge that has an associated direction shown by an arrow. A **digraph** is a graph in which the edges are directed.
- A **path** is a finite sequence of edges such that the end vertex of one edge is the start vertex of the next edge. No edge is included more than once.
- A **connected graph** is one in which every pair of vertices is connected by a path.
- A **complete graph** is one in which every pair of vertices is connected by an edge. A complete graph with n vertices is denoted by K_n.
- A **planar graph** is one that can be drawn in a plane such that no two edges meet except at a vertex. K_4 is planar but K_5 is not planar.

K_4

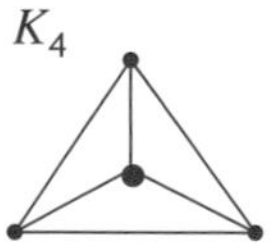

K_5

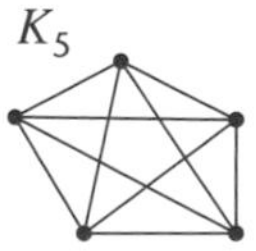

- A **cycle** is a path that starts and finishes at the same vertex.
- An **Eulerian** cycle is one that traverses all of the edges of the graph.
- A **tree** is a graph with no cycles.
- A **spanning tree** is a tree whose vertices are all of the vertices of the graph.
- A **network** is a graph that has a number (**weight**) associated with every edge.
- A **minimum spanning tree** (MST) of a network is a spanning tree of minimum possible weight. It is sometimes called a **minimum connector**.

Prim's algorithm

Prim's algorithm may be used to find a minimum spanning tree of a network.

Step 1 Choose a starting vertex.

Step 2 Connect it to the vertex that will make an edge of minimum weight.

Step 3 Connect one of the remaining vertices to the tree formed so far, in such a way that the minimum extra weight is added to the tree.

Step 4 Repeat step 3 until all of the vertices are connected.

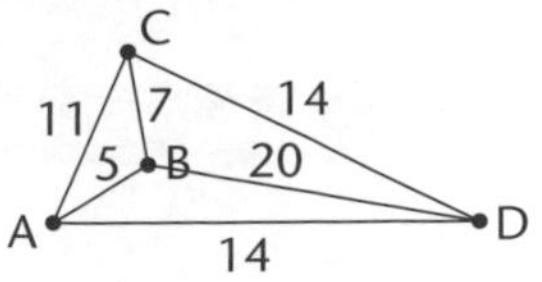

Starting from A gives the MST as

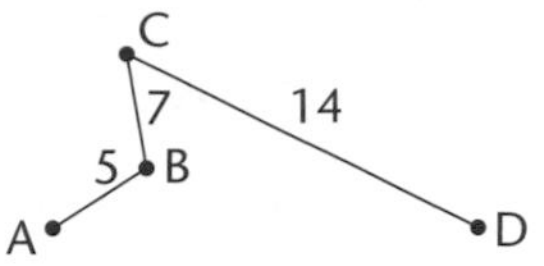

AD could have been used instead of CD

The number in row A and column B, for example, represents the weight of the edge AB.

The network given above may be represented by the matrix shown.

	A	B	C	D
A	–	5	–	–
B	5	–	7	–
C	–	7	–	14
D	–	–	14	–

A version of Prim's algorithm may be used to find the minimum spanning tree directly from a matrix.

Step 1 Choose a starting vertex. Delete the corresponding row and write 1 above the corresponding column as a label.

Step 2 Circle the smallest undeleted value in the labelled column and delete the row in which it lies.

Step 3 Label the column, corresponding to the vertex of the deleted row, with the next label number.

If there is more than one smallest value then you can choose which one to circle.

Step 4 Circle the smallest undeleted value of all the values in the labelled columns and delete the row in which it lies.

Step 5 Repeat steps 3 and 4 until all of the rows are deleted. The circled values then define the edges of the minimum spanning tree.

	1	2	3	
	A	B	C	D
~~A~~	~~–~~	~~5~~	~~–~~	~~–~~
~~B~~	~~(5)~~	~~–~~	~~7~~	~~–~~
~~C~~	~~–~~	~~(7)~~	~~–~~	~~14~~
~~D~~	~~–~~	~~–~~	~~(14)~~	~~–~~

The circled values correspond to the edges AB, BC, and CD.

Kruskal's algorithm

Kruskal's algorithm is an alternative way to find a minimum spanning tree. In this case, the subgraph produced may not be connected until the final stage.

Step 1 Choose an edge with minimum weight as the first subgraph.

Step 2 Find the next edge of minimum weight that will not complete a cycle when taken with the existing subgraph. Include this edge as part of a new subgraph.

Step 3 Repeat step 2 until the subgraph makes a spanning tree.

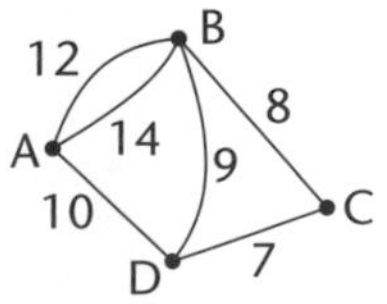

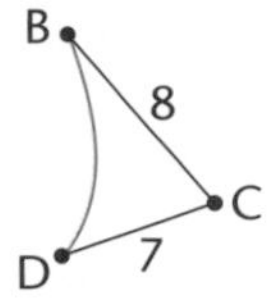

BD can not be included

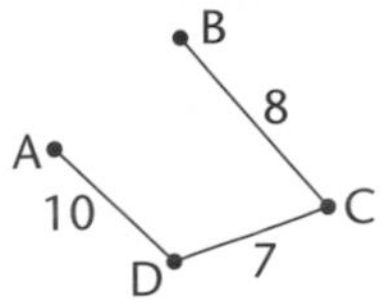

Dijkstra's algorithm

The word *distance* is used here in place of the more general *weight* because in practical applications of Dijkstra's algorithm the weight so often represents a distance.

Reading the letters inside the box as OWL might help you to remember which part is for which information.

Dijkstra's algorithm is used to find the shortest distance between a chosen *start* vertex and any other vertex in a network.

The algorithm is designed for use by computers but, when implementing it without the aid of a computer, a system of labelling is required.

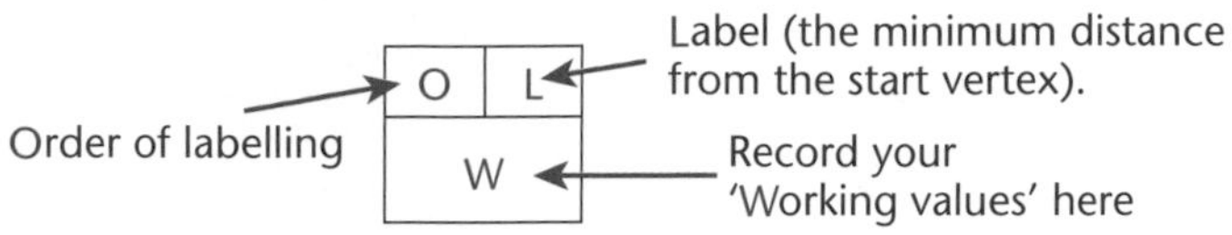

The steps in the algorithm below refer to this system of labelling. Every vertex is labelled in the same way.

Step 1 Give the start vertex a label of 0 and write its order of labelling as 1.

Step 2 For each vertex directly connected to the start vertex, enter the distance from the start vertex as a working value. Enter the smallest working value as a label for that vertex. Record the order in which it was labelled as 2.

Step 3 For each vertex directly connected to the one that was last given a label, add the edge distance onto the label value to obtain a total distance. This distance becomes the working value for the vertex unless a lower one has already been found.

Step 4 Find the vertex with the smallest working value not yet labelled and label it. Record the order in which it was labelled.

Step 5 Repeat steps 3 and 4 until the target vertex is labelled. The value of the label is the minimum distance from the start vertex.

Step 6 To find the path that gives the shortest distance, start at the target vertex and work back towards the start vertex in such a way that an edge is only included if its distance equals the change in the label values.

Example

Use Dijkstra's algorithm to find the shortest distance from A to F in this network and state the route used.

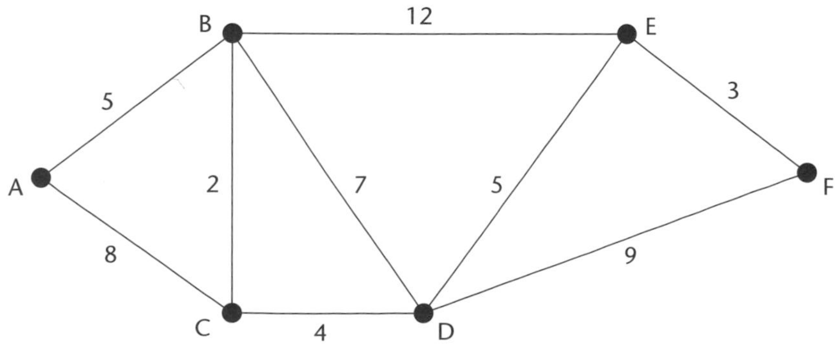

Using Dijkstra's algorithm produces the diagram below.

The shortest distance from A to F is 19 (given by the label at F).

Tracing the route backwards using Step 6 gives F, E, D, C, B, A, so the required route is A, B, C, D, E, F.

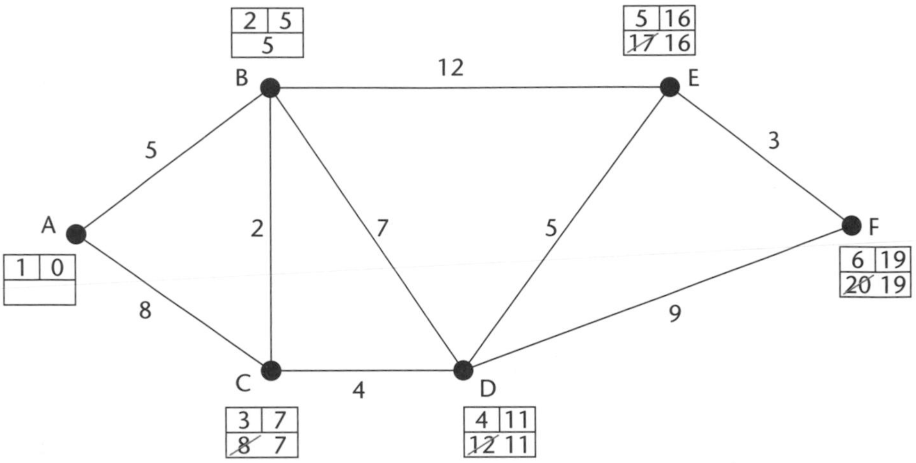

The route inspection problem

Again, it is common to use length rather than weight in this context.

The **route inspection problem** is to find a route of minimum total length that traverses every edge of the network, at least once, and returns to the start vertex.

This is also known as the Chinese postman problem.

Leonhard Euler was the most prolific mathematician of all time. One of his many contributions to the subject was to introduce graph theory as a means of solving problems.

The algorithm for solving this problem is based on the idea of a **traversable** graph. A graph is traversable if it can be drawn in one continuous movement without going over the same edge more than once. If all of the vertices of the graph are even, then any one of them may be used as the starting point and the same point will be returned to at the end of the movement. Such a graph is said to be **Eulerian.** The only other possibility for a traversable graph is that it has exactly two odd vertices. In this situation, one of the vertices is the start point and the other is the finish point. This type of graph is said to be **Semi-Eulerian.**

The **route inspection algorithm** is:

Step 1 List all of the odd vertices.

Step 2 Form the list into a set of pairs of odd vertices. Find all such sets.

Step 3 Choose a set. For each pair find a path of minimum length that joins them. Find the total length of these paths for the chosen set.

Step 4 Repeat steps 3 until all sets have been considered.

Step 5 Choose the set that gives the minimum total. Each pair in the set defines an edge that must be repeated in order to solve the problem.

Repeating the edges between the pairs of odd vertices in this way effectively makes all the vertices even and creates a traversable graph.

Progress check

1 (a) Use Prim's algorithm to find the total weight of a minimum spanning tree for this network.

(b) Verify your answer to part (a) using Kruskal's algorithm.

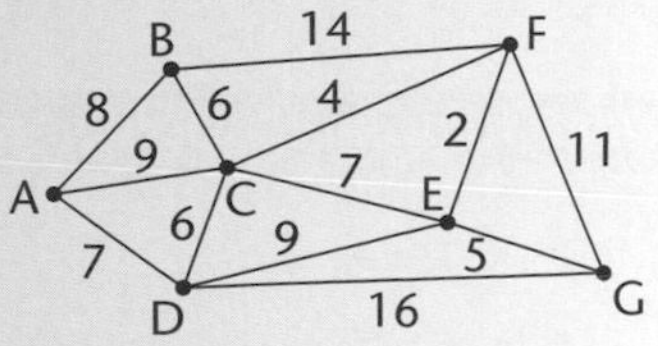

2 Use Dijkstra's algorithm to find the shortest distance from A to G in the network given in question 1. State the route used.

1 (a) 30

2 20, ACFEG

5.3 Critical path analysis

After studying this section you should be able to:

- construct an activity network from a given precedence table including the use of dummies where necessary
- use forward and backward scans to determine earliest and latest event times
- find the critical path in an activity network
- calculate the total float of an activity
- construct a chart for the purpose of scheduling

LEARNING SUMMARY

The process of representing a complex project by a network and using it to identify the most efficient way to manage its completion is called **critical path analysis.**

Activity networks

A complex project may be divided into a number of smaller parts called **activities.** The completion of one or more activities is called an **event.**

Activities often rely on the completion of others before they can be started.

The relationship between these activities can be represented in a **precedence table**, sometimes called a **dependency table.**

In the precedence table shown on the right, the figures in brackets represent the **duration** of each activity, i.e. the time required, in hours, for its completion.

Activity	Depends on
A(3)	–
B(5)	–
C(2)	A
D(3)	A
E(3)	B, D
F(5)	C, E
G(1)	C
H(2)	F, G

A precedence table can be used to produce an **activity network**. In the network, activities are represented by arcs and events are represented by vertices.

The vertices are numbered from 0 at the **start vertex** and finishing at the **terminal vertex.**

The direction of the arrows shows the order in which the activities must be completed.

There must only be *one* activity between each pair of events in the network. The notation (i, j) is used to represent the activity between events i and j.

A **dummy activity** is one that has zero duration. A dummy is needed in this network to show that G depends on C whereas F depends on C *and* E.

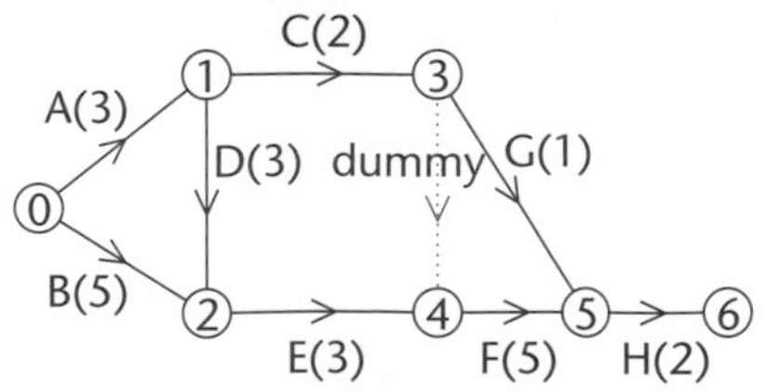

A dummy is shown with a dotted line. Its direction is important in defining dependency. In this case it shows that F depends on C not that G depends on E.

Earliest event times

The **earliest event time** for vertex i is denoted by e_i and represents the earliest time of arrival at event i with all dependent activities completed. These times are calculated using a **forward scan** from the start vertex to the terminal vertex.

Latest event times

The **latest event time** for vertex i is denoted by l_i and represents the latest time that event i may be left without extending the time for the project. These times are calculated using a **backward scan** from the terminal vertex back to the start vertex.

The **critical path** is the longest path through the network. The activities on this path are the **critical activities.** If any critical activity is delayed then this will increase the time needed to complete the project. The events on the critical path are the **critical events** and for each of these $e_i = l_i$.

It is useful to add the information about earliest and latest times to the network. The critical path is then easily identified.

Notice that B(5) lies between two critical events but it is not a critical activity.

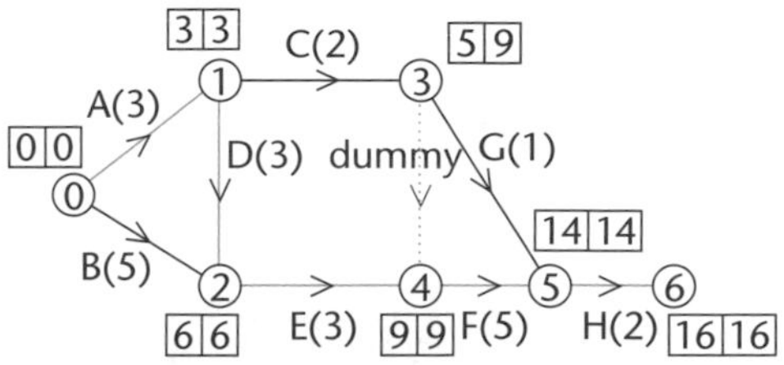

The total float of any critical activity is always zero.

The **total float of an activity** is the maximum time that the activity may be delayed without affecting the length of the critical path. It is given by:

latest finish time – earliest start time – duration of the activity.

Scheduling

The process of allocating activities to workers for completion, within all of the constraints of the project, is known as **scheduling**.

The information regarding earliest and latest times for each activity is crucial when constructing a schedule. This information may be presented as a table or as a chart.

Typically, the purpose of scheduling is to determine the number of workers needed to complete the project in a given time, or to determine the minimum time required for a given number of workers to complete the project.

		Start		Finish		
Activity	Duration	Earliest	Latest	Earliest	Latest	Float
A(0, 1)	3	0	0	3	3	0
B(0, 2)	5	0	1	5	6	1
C(1, 3)	2	3	7	5	9	4
D(1, 2)	3	3	3	6	6	0
E(2, 4)	3	6	6	9	9	0
F(4, 5)	5	9	9	14	14	0
G(3, 5)	1	5	13	6	14	8
H(5, 6)	2	14	14	16	16	0

The critical activities are shown along one line.

This type of diagram is called a Gantt chart or a cascade chart.

The diagram illustrates the degree of flexibility in starting activities B, C and G. Remember that G cannot be started until C has been completed.

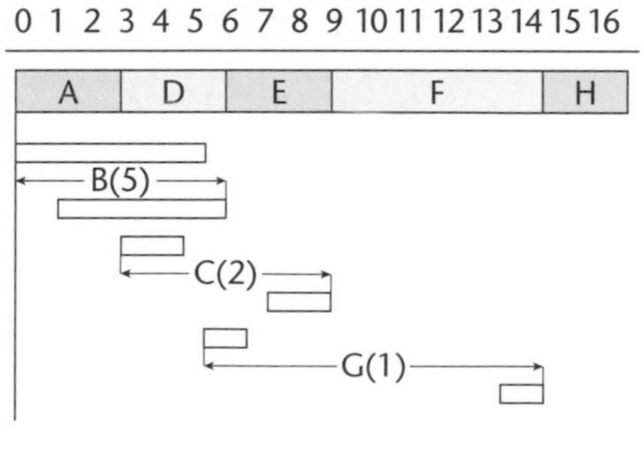

Progress check

1 Determine the critical activities and the length of the critical path for this network.

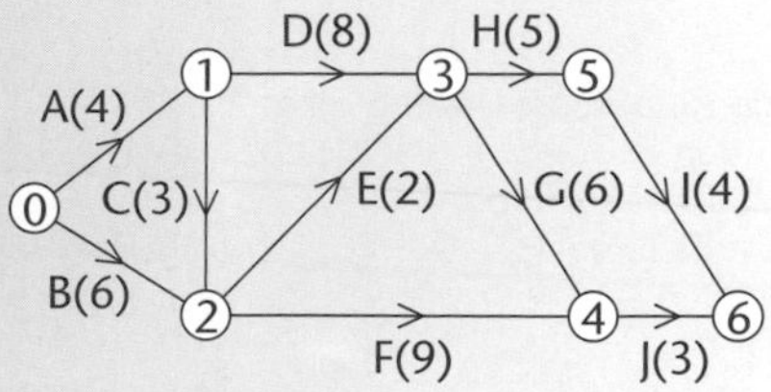

1 A, D, G and J. 21

5.4 Linear programming

After studying this section you should be able to:

- *formulate a linear programming problem in terms of decision variables*
- *use a graphical method to represent the constraints and solve the problem*

LEARNING SUMMARY

Formulating a linear programming problem

To formulate a linear programming problem you need to:

- Identify the **variables** in the problem and give each one a label.
- Express the **constraints** of the problem in terms of the variables. You need to include non-negativity constraints such as $x \geqslant 0$, $y \geqslant 0$.
- Express the quantity to be optimised in terms of the variables. The expression produced is called the **objective function.**

x and y are often used for the variables.

Typically, this may be to maximise a profit or minimise a loss.

Example

A small company produces two types of armchair. The cost of labour and materials for the two types is shown in the table.

	Labour	Materials
Standard	£30	£25
Deluxe	£40	£50

The total spent on labour must not be more than £1150 and the total spent on materials must not be more than £1250. The profit on a standard chair is £70 and the profit on a deluxe chair is £100.
How many chairs of each type should be made to maximise the profit?

In this case, the variables are the number of chairs of each type that may be produced. Using x to represent the number of standard chairs and y to represent the number of deluxe chairs, the constraints may be written as:

$$30x + 40y \leqslant 1150 \Rightarrow 3x + 4y \leqslant 115$$

$$25x + 50y \leqslant 1250 \Rightarrow x + 2y \leqslant 50$$

and $x \geqslant 0, \quad y \geqslant 0.$

It's a good idea to simplify the constraints where possible.

Using P to stand for the profit, the problem is to maximise $P = 70x + 100y$.

The graphical method of solution

Each constraint is represented by a region on the graph. It's a good idea to shade out the *unwanted* region for each one. The part that remains unshaded then defines the **feasible region** containing the points that satisfy all of the constraints.

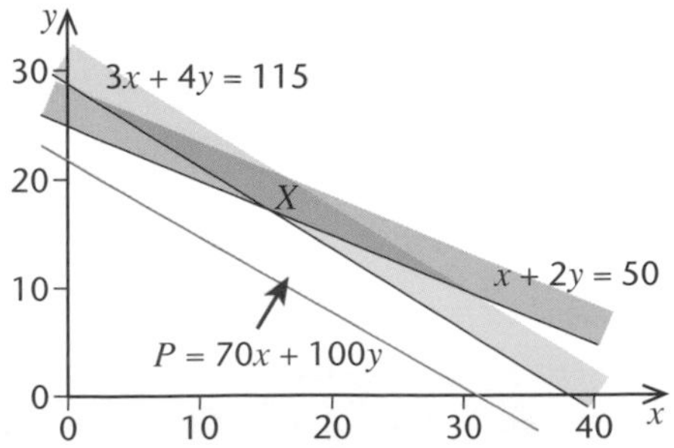

The blue line represents the points where the profit takes a particular value. Moving the line in the direction of the arrow corresponds to increasing the profit. This suggests that the maximum profit occurs at the point X.

Solving $3x + 4y = 115$ and $x + 2y = 50$ simultaneously gives X as (15, 17.5).

X does not represent the solution in this case because both x and y must be integers.

The nearest points with integer coordinates in the feasible region are (15,17) and (14,18) The profit, given by $P = 70x + 100y$, is greater at (14,18).

The maximum profit is made by producing 14 standard and 18 deluxe chairs.

Progress check

1 In the diagram, the shaded region represents the feasible region for a linear programming problem. All of the boundary lines are included.

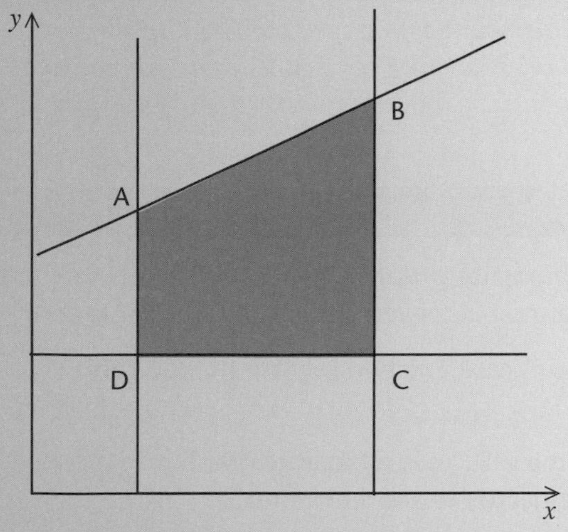

The object is to maximise a profit given by $P = 1.2x + 0.9y$

Which point of the feasible region represents the maximum profit?

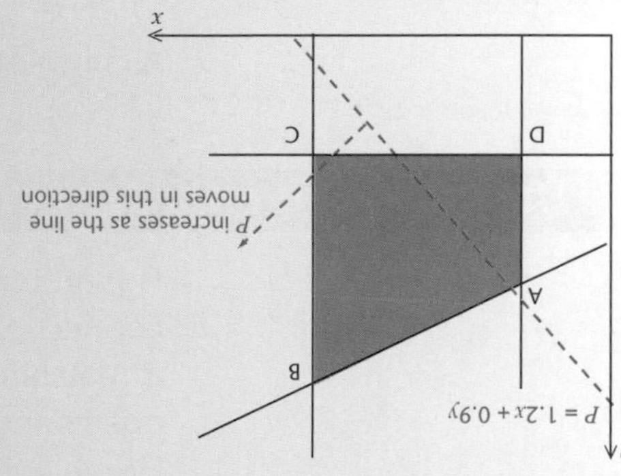

1 B (see diagram for explanation)

5.5 Matchings

After studying this section you should be able to:

- *use bipartite graphs to model matchings*
- *understand the conditions for matchings to be maximal or complete*
- *apply the maximum matching algorithm*

LEARNING SUMMARY

Matchings and graphs

A **bipartite graph** is a graph in which the vertices are divided into two sets such that no pair of vertices in the same set is connected by an edge.

In this case, the two sets are {A, B, C} and {p, q, r, s}.

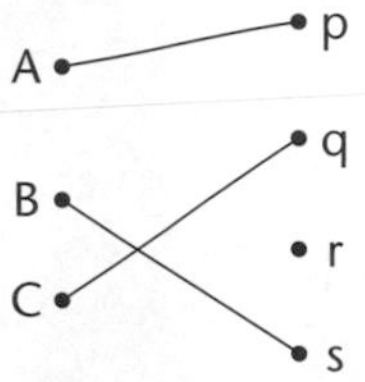

Some vertices in a bipartite graph may not be connected to another vertex.

A **matching** between two sets may be represented by a bipartite graph in which there is at most one edge connecting a pair of vertices.

A **maximal matching** is a matching which has the maximum number of edges. This occurs when every vertex in one of the sets is connected to a vertex in the other set. The bipartite graph shown above represents a maximal matching.

A **complete matching** is a matching in which every vertex is connected to another vertex. This can only occur when the two sets contain the same number of vertices.

The matching improvement algorithm

Figure 1 is a bipartite graph showing the possible connections between two sets. It does not represent a matching because some vertices have more than one connection.

Figure 2 is a bipartite graph representing an initial matching.

An initial matching may be improved by increasing the number of connections. This is the purpose of the **matching improvement algorithm**.

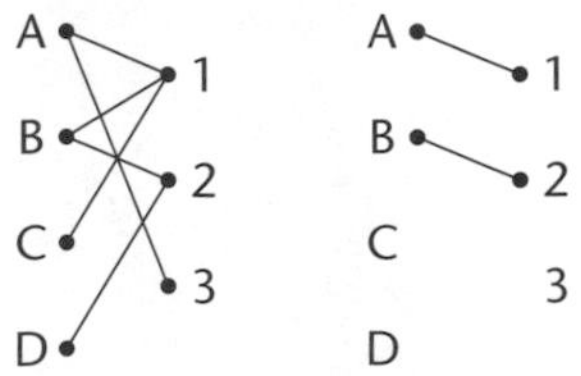

Figure 1 Figure 2

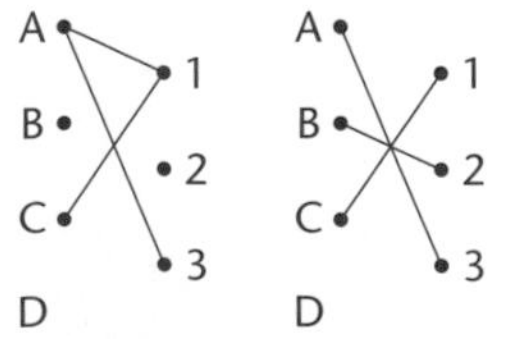

Alternating path Maximal matching

In an alternating path, the edges alternate between those that are not in the initial matching and those that are.

Step 1 In the initial matching, start from a vertex that is not connected and look for an **alternating path** to a vertex in the other set that is not connected.

Step 2 Each edge on the alternating path, not included in the initial matching, is now included and each edge originally included is removed.

Step 3 Repeat steps 1 and 2 using the latest matching in place of the initial matching until no further alternating paths can be found.

Progress check

1 Starting with an initial matching in which 1 is connected to R and 2 is connected to P, use the improvement algorithm to establish a complete matching for this bipartite graph.

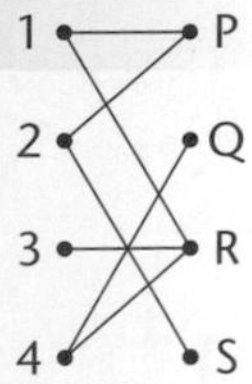

1 1–P, 2–S, 3–R, 4–Q.

Sample questions and model answers

1

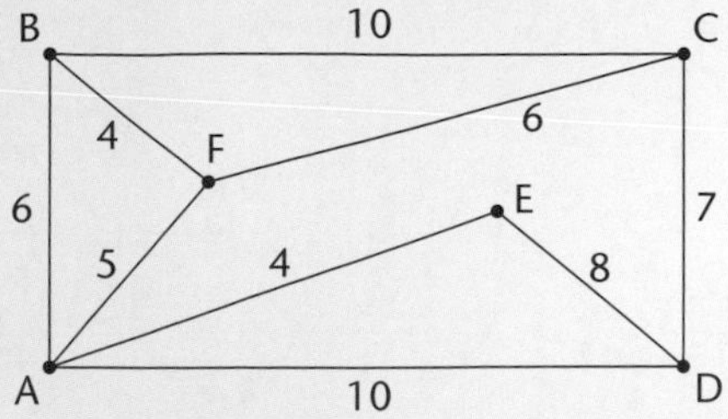

The diagram represents the system of pathways that a security guard must patrol during his course of duty. The weights on the edges represent the time taken, in minutes, to patrol each pathway.

Find the minimum time required to patrol every pathway at least once and give a possible route.

> You need to recognise that this question requires the use of the route inspection algorithm.

Vertex	Order
A	4
B	3
C	3
D	3
E	2
F	3

> Step 1 of the algorithm is to list the odd vertices.

The odd vertices are B, C, D and F.

The sets of pairs of odd vertices are: {BC, DF}
{BD, CF}
{BF, CD}.

> Step 2 of the algorithm is to find the sets of pairs of odd vertices. Each odd vertex appears once in each set.

For the set {BC, DF}

BC = 10. The shortest route from D to F is DC + CF = 13.

Total length of the extra paths for this set is 10 + 13 = 23.

> There is a an edge connecting B and C.

> Step 3 of the algorithm is to find the extra length introduced for each set.

For the set {BD, CF}

The shortest route from B to D is BA + AD = 16. CF = 6.

Total length of the extra paths for this set is 16 + 6 = 22.

For the set {BF, CD}

> There is an edge connecting C and F.

BF = 4 and CD = 7.

Total length of the extra paths for this set is 4 + 7 = 11.

The minimum total corresponds to the set {BF, CD} which means that the edges BF and CD will be repeated.

> Step 4 of the algorithm is to keep going until all the sets have been considered.

A possible route is ABFCDEAFBCDA.

> There are many possible routes.

The time taken will be 60 + 11 = 71 minutes.

Sample questions and model answers (continued)

2

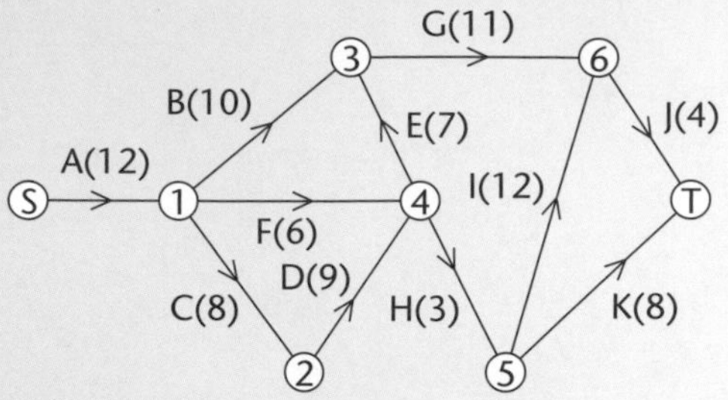

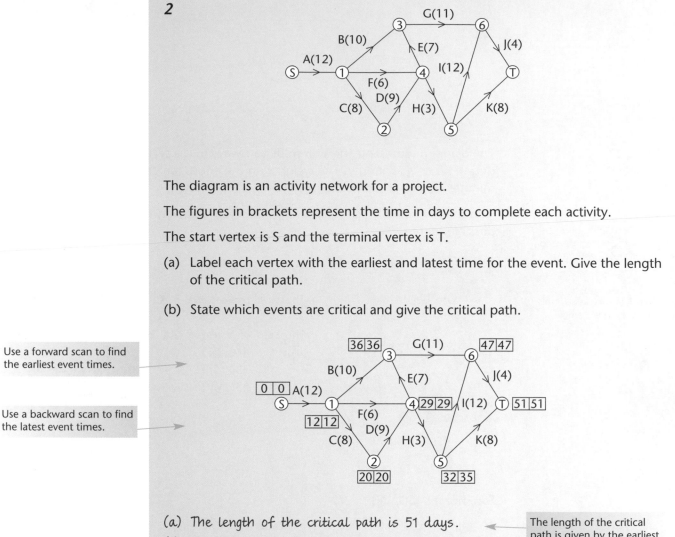

The diagram is an activity network for a project.

The figures in brackets represent the time in days to complete each activity.

The start vertex is S and the terminal vertex is T.

(a) Label each vertex with the earliest and latest time for the event. Give the length of the critical path.

(b) State which events are critical and give the critical path.

Use a forward scan to find the earliest event times.

Use a backward scan to find the latest event times.

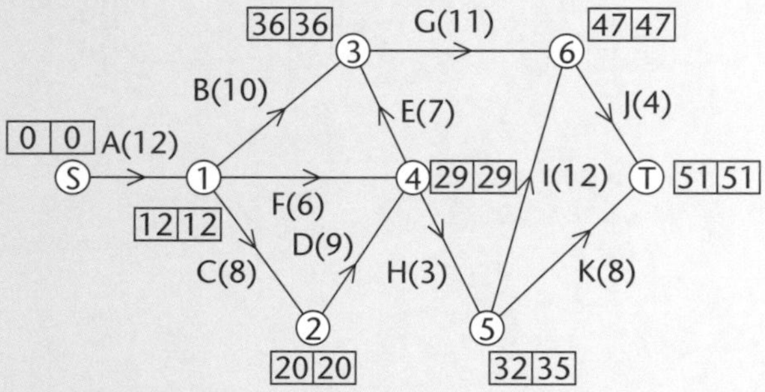

(a) The length of the critical path is 51 days.

(b) The critical events are S, 1, 2, 3, 4, 6 and T.

The critical path is A, C, D, E, G and J.

The length of the critical path is given by the earliest (and latest) event time at the terminal vertex.

Practice examination questions

1 The table shows the amount of memory taken up by some files on a computer system.

File	A	B	C	D	E	F	G
Memory (Kb)	580	468	610	532	840	590	900

The files are to be transferred onto disks with a capacity of 1.4 Mb (1000 Kb = 1 Mb).

(a) Show that a minimum of four disks is required.

(b) Use the first-fit algorithm to allocate the files to disks.

(c) Use the first-fit decreasing algorithm to show how all of the files may be stored using four disks.

2 A new theme park has its major attractions at A, B, C, D, E, F and G as shown in the network below. The edges of the network show the *possible* routes of paths connecting the attractions.

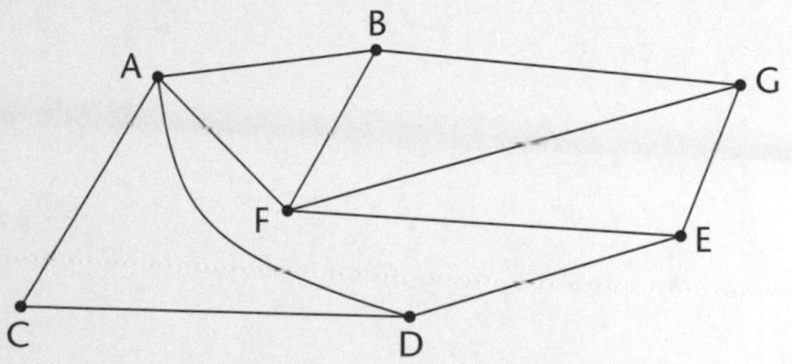

The cost of laying each of these paths, in hundreds of pounds, is given in the table:

	A	B	C	D	E	F	G
A	–	50	40	61	–	38	–
B	50	–	–	–	–	47	54
C	40	–	–	35	–	–	–
D	61	–	35	–	42	–	–
E	–	–	–	42	–	25	10
F	38	47	–	–	25	–	63
G	–	54	–	–	10	63	–

The paths that are actually laid must form a connected network.
Use Prim's algorithm, starting by deleting row A, to find the minimum cost of laying the necessary paths.

Practice examination questions (continued)

3

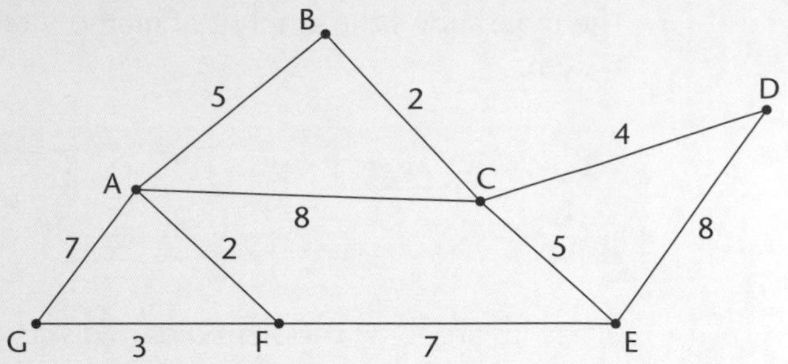

(a) Use Dijkstra's algorithm to find the shortest distance from G to D.

(b) Describe the route taken.

4

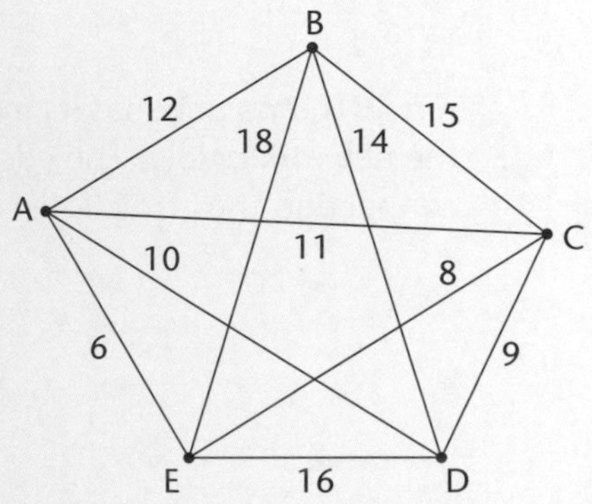

(a) Use Kruskal's algorithm to find the length of the minimum spanning tree for this network.

State the order in which you include each of the edges.

(b) Find an upper bound for the travelling salesman problem using the minimum spanning tree with shortcuts.

5

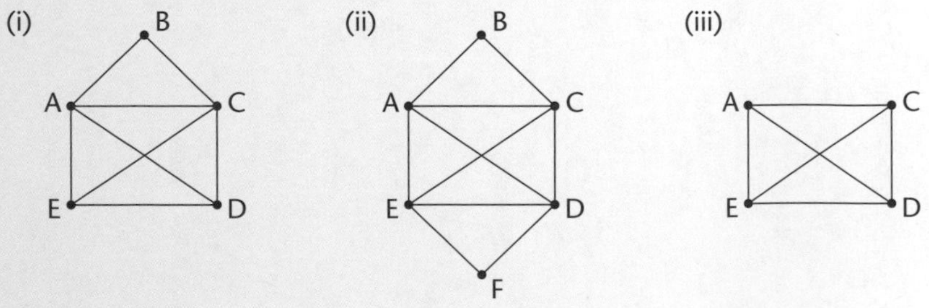

List the order of the vertices for each graph.

Practice examination questions *(continued)*

6 A distribution manager has the task of transporting pallets of frozen food between two storage centres. The maximum number of pallets is to be delivered within a daily budget of £2000. Three types of van are available and each van can only do the trip once per day.

The details of what the vans can carry and the daily costs are shown in the table.

At most 10 drivers may be used in one day.

Van type	No. of pallets	Cost/day
Class A	4	£120
Class B	9	£300
Class C	12	£400

Let a represent the number of Class A vans used, b the number of Class B vans used, c the number of Class C vans used and P the total number of pallets transported in a day.

(a) Write an expression for the objective function.

(b) Formulate the task as a linear programming problem.

(c) Describe any special condition that the solution must satisfy.

7 The table shows the names of five employees of a company and the days when they are each available to work a late shift. The days that have been highlighted represent a first attempt at matching the employees to the available days to make a roster.

(a) Use a bipartite graph to represent the relationship between employees and days available.

(b) Use a second bipartite graph to represent an initial matching based on the days highlighted.

(c) Implement a matching improvement algorithm to obtain a complete matching. Indicate the alternating path used.

Name	Days available
Abbie	**Mon**, Tue, Fri
Ben	Mon, **Wed**, Thu
Colin	Mon, **Fri**
Dave	Fri, Wed
Emma	**Thu**, Fri

Practice examination answers

Core 1

1 $49^{-\frac{1}{2}} = \frac{1}{49^{\frac{1}{2}}} = \frac{1}{\sqrt{49}} = \frac{1}{7}$.

2 $8^{x-3} = 4^{x+1} \Rightarrow (2^3)^{x-3} = (2^2)^{x+1}$

$\Rightarrow 2^{3(x-3)} = 2^{2(x+1)}$

$\Rightarrow 3(x-3) = 2(x+1)$

$\Rightarrow x = 11$.

3 (a) $x^2 - 4x + 1 = (x-2)^2 - 3$.

(b) $(x-2)^2 - 3 = 0$

$\Rightarrow x - 2 = \pm\sqrt{3}$

$\Rightarrow x = 2 \pm \sqrt{3}$.

(c) (2, –3).

(d)

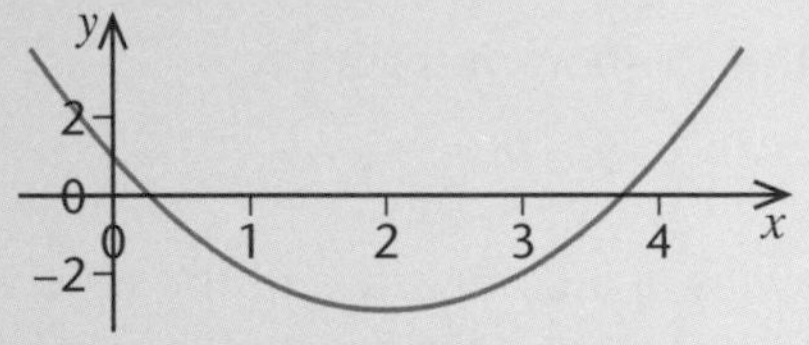

4 (a) (b)

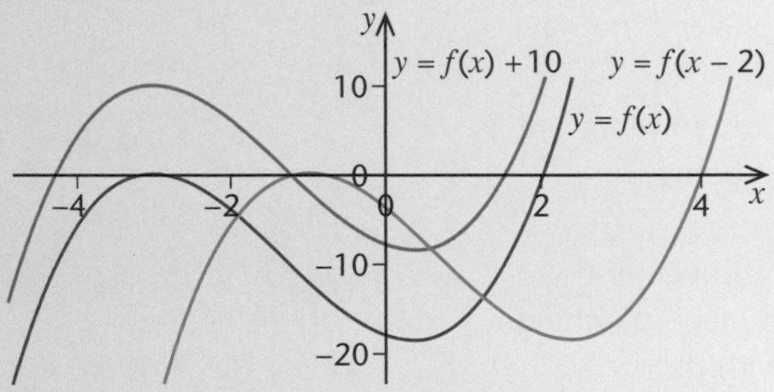

5 The coordinates at the points of intersection are given by the simultaneous solution of:

$y = 2x - 2$ [1]

$y = x^2 - x - 6$ [2]

Eliminating y gives

$x^2 - x - 6 = 2x - 2$

$\Rightarrow \quad x^2 - 3x - 4 = 0$

$\Rightarrow \quad (x+1)(x-4) = 0$

$\Rightarrow \quad x = -1$ or $x = 4$.

When $x = -1$, $y = -4$.

When $x = 4$, $y = 6$.

The points of intersection are A(–1,–4) and B(4, 6).

6 $3x^2 - 8x - 7 < 2x^2 - 3x - 11$

$\Rightarrow x^2 - 5x + 4 < 0$

$\Rightarrow (x-1)(x-4) < 0$

$\Rightarrow 1 < x < 4$.

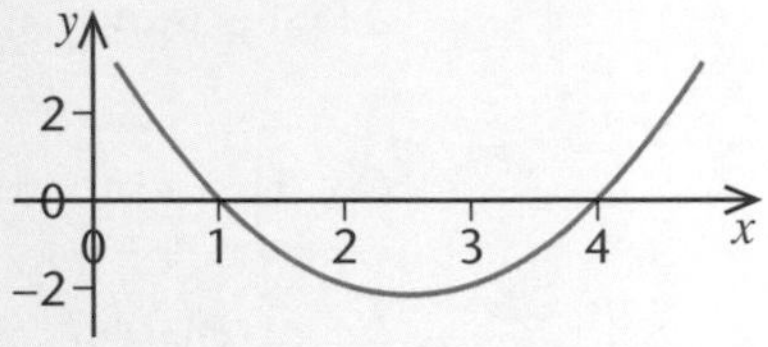

7 (a) Gradient of AB = 6/5.

Using $y - y_1 = m(x - x_1)$ gives

$y - 1 = 6/5(x - 4)$

$\Rightarrow \quad 5y - 5 = 6x - 24$

$\Rightarrow \quad 6x - 5y - 19 = 0$.

(b) The mid point of A and B is (1.5, –2).

The gradient of the line l is –5/6.

Using $y - y_1 = m(x - x_1)$ again gives

$y + 2 = -5/6(x - 1.5)$

$\Rightarrow \quad 6y + 12 = -5x + 7.5$

$\Rightarrow \quad 12y + 24 = -10x + 15$

$\Rightarrow \quad 10x + 12y + 9 = 0$.

(c) At C, $x = 0 \Rightarrow y = -0.75$ so the coordinates of C are (0, –0.75).

Core 1 (continued)

8 (a) The equation of the line m is $3x + 2y = 6$

$\Rightarrow$ $y = -3/2x + 3$

$\Rightarrow$ The gradient of the line m is $-3/2$

$\Rightarrow$ The gradient of the line l is $2/3$.

Using $y - y_1 = m(x - x_1)$ gives

$y - 4 = 2/3(x - 5)$

$\Rightarrow$ $3y - 12 = 2x - 10$

$\Rightarrow$ $2x - 3y + 2 = 0$ which is the equation of line l.

(b) At A, $y = 0 \Rightarrow x = -1$.

B lies on the line with equation $3x + 2y = 6$.

At B, $y = 0 \Rightarrow x = 2$.

So the distance AB is given by $2 - (-1) = 3$.

9 10th term $= a + 9d = 74$. [1]

Sum of first 20 terms $= 10(2a + 19d) = 1510$

$\Rightarrow$ $2a + 19d = 151$ [2]

$2 \times$ [1] gives $2a + 18d = 148$ [3]

[2] – [3] gives $d = 3$

Subs for d in [1] gives $a + 27 = 74 \Rightarrow a = 47$.

First term = 47, common difference = 3.

10 (a) $x + y = 7$

(b) At P and Q $x + y = 7$ [1]

and $y = (x - 1)^2 + 4$ [2]

Substituting for y in [2] gives

$7 - x = x^2 - 2x + 5$

$\Rightarrow x^2 - x - 2 = 0$

$\Rightarrow (x - 2)(x + 1) = 0$

$\Rightarrow x = 2$ or $x = -1$.

$x = 2 \Rightarrow y = 5$ and $x = -1 \Rightarrow y = 8$.

P is the point (–1, 8) and Q is the point (2, 5).

Core 2

1 (a) First term $a = 3$, common ratio $r = 2$.
The 8th term is $ar^7 = 3 \times 2^7 = 384$.

(b) First term $a = 25$, common difference $d = 1.2$

number of terms $n = 50$.

Using $S_n = \frac{n}{2}(2a + (n - 1)d)$ gives

Sum $= \frac{50}{2}(2 \times 25 + 49 \times 1.2) = 2720$.

2 (a) By the factor theorem $f(-2) = 0$

$\Rightarrow (-2)^3 - 4(-2)^2 - 3(-2) + k = 0$

$\Rightarrow -8 - 16 + 6 + k = 0$

$\Rightarrow k = 18$.

(b) $f(x) = x^3 - 4x^2 - 3x + 18$

$\equiv (x + 2)(Ax^2 + Bx + C)$.

Comparing coefficients of x^3 gives A = 1.

Comparing the constant terms gives C = 9.

When $x = -1$

$-1 - 4 + 3 + 18 = A - B + C$

$\Rightarrow 16 = 1 - B + 9 \Rightarrow B = -6$.

$f(x) = (x + 2)(x^2 - 6x + 9)$

$= (x + 2)(x - 3)^2$.

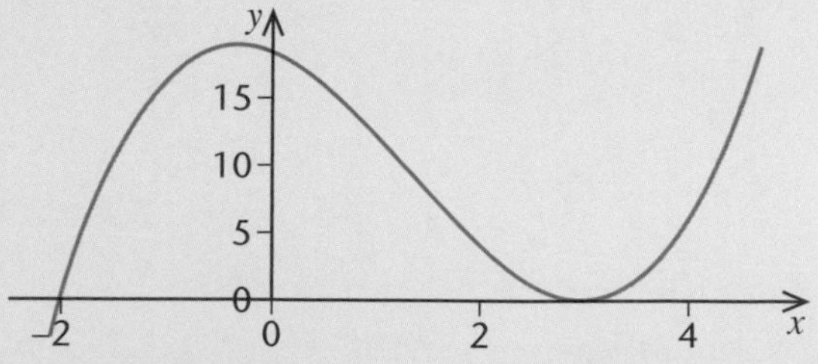

(c) From the sketch, $f(x) \geqslant 0 \Rightarrow x \geqslant -2$.

(d) By the remainder theorem, the remainder is given by $f(-1) = 16$.

Core 2 (continued)

3 (a) $3\log_a x - 2\log_a y + \log_a(x+1)$

$= \log_a x^3 - \log_a y^2 + \log_a(x+1)$

$= \log_a\left(\dfrac{x^3(x+1)}{y^2}\right).$

(b) $5^x = 100$

$\Rightarrow \log_{10} 5^x = \log_{10} 100$

$x \log_{10} 5 = \log_{10} 100$

$x = \dfrac{\log_{10} 100}{\log_{10} 5} = 2.861$ to 3 d.p.

4 $\cos^{-1}(0.4) = 66.42\ldots°$

$0° \leqslant x \leqslant 360° \Rightarrow 30° \leqslant 2x + 30 \leqslant 750°$

Sketch the graph of $y = \cos X$, $30° \leqslant X \leqslant 750°$,

and the graph of $y = 0.4$

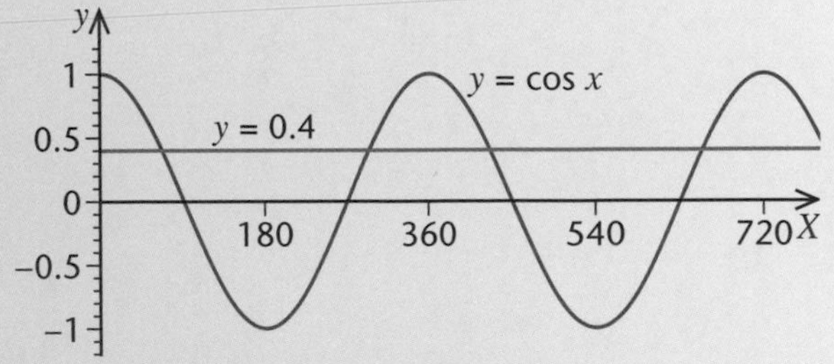

From the sketch, the values of X (i.e $2x + 30$) are:

$66.42\ldots°$, $360 - 66.42\ldots°$, $360 + 66.42\ldots°$, $720 - 66.42\ldots°$,

This gives

$x = 18.2°$ or $x = 131.8°$ or $x = 198.2°$ or $x = 311.8°$.

5 (a) $\cos x + 3\sin x \tan x - 2 = 0$

$\Rightarrow \cos^2 x + 3\sin^2 x - 2\cos x = 0$

$\Rightarrow \cos^2 x + 3(1 - \cos^2 x) - 2\cos x = 0$

$\Rightarrow 3 - 2\cos^2 x - 2\cos x = 0$

$\Rightarrow 2\cos^2 x + 2\cos x - 3 = 0$

(b) $\cos x = \dfrac{-2 \pm \sqrt{4 + 24}}{4}$

$\Rightarrow \cos x = 0.8228\ldots$ or $\cos x = -1.8228\ldots$ (No solutions)

$\Rightarrow x = 34.6°$ or $x = 325.4°$

6 (a) $f'(x) = 2x - x^{-2}$

$f'(2) = 4 - \frac{1}{4} = 3\frac{3}{4}$

(b) $f''(x) = 2 + 2x^{-3}$

$f''(-1) = 2 - 2 = 0.$

7 Area $= \displaystyle\int_1^4 3x^{\frac{1}{2}}\,dx$

$= [2x^{\frac{3}{2}}]_1^4 = 2(8 - 1)$

$= 14.$

8 Area sector = area triangle + area segment

$\Rightarrow \dfrac{1}{2}\theta = \dfrac{1}{2}\sin\theta + \dfrac{\pi}{5}$

$\Rightarrow \theta = \sin\theta + \dfrac{2\pi}{5}.$

9 Area $= \displaystyle\int_2^3 \sqrt{x^2 - 3}\,dx.$

$d = \dfrac{3-2}{5} = 0.2$ Taking $f(x) = \sqrt{x^2 - 3}$ gives

$A = 0.1(f(2) + 2(f(2.2) + f(2.4) + f(2.6) + f(2.8)) + f(3))$

$A = 0.1(1 + 2(1.3565 + 1.6613 + 1.9591 + 2.2) + 2.4495)$

$A = 1.78$ to 2 d.p.

10 (a) $2x^{\frac{3}{2}}\left(1 - \dfrac{x}{5}\right)$

(b) (0, 0), (5, 0)

(c) $\dfrac{dy}{dx} = 3x^{\frac{1}{2}} - x^{\frac{3}{2}}$

(d) $\dfrac{dy}{dx} = 0 \Rightarrow x^{\frac{1}{2}}(3 - x) = 0$

$\Rightarrow x = 0$ or $x = 3$

Since $x > 0$, $x = 3$

when $x = 3$, $y = 4.157$ (4 s.f.)

$\dfrac{d^2y}{dx^2} = \frac{3}{2}x^{-\frac{1}{2}} - \frac{3}{2}x^{\frac{1}{2}}$

when $x = 3$, $\dfrac{d^2y}{dx^2} = \frac{3}{2}(3^{-\frac{1}{2}} - 3^{\frac{1}{2}}) < 0$

So the turning point at (3, 4.157) is a maximum.

11 (a) $y = 2x^3 - 15x^2 - 36x + 10$

$\dfrac{dy}{dx} = 6x^2 - 30x - 36$

At a stationary point

$\dfrac{dy}{dx} = 0 \Rightarrow 6x^2 - 30x - 36 = 0$

$\Rightarrow x^2 - 5x - 6 = 0$

$\Rightarrow (x + 1)(x - 6) = 0$

$\Rightarrow x = -1$ or $x = 6$.

When $x = -1$, $y = 29$.

When $x = 6$, $y = -314$.

The stationary points are $(-1, 29)$ and $(6, -314)$.

(b) $\dfrac{d^2y}{dx^2} = 12x - 30$

When $x = -1$, $\dfrac{d^2y}{dx^2} = -42 < 0$ maximum.

When $x = 6$, $\dfrac{d^2y}{dx^2} = 42 > 0$ minimum.

(c) The function is decreasing at all of the points between the turning points i.e. when $-1 < x < 6$.

Mechanics 1

1 (a) $u = 10$

$a = -4$

Using $v = u + at$ $\quad v = 10 - 4 \times 5 = -10.$

The speed of the particle when $t = 5$ is 10 m s^{-1}.

(b) $v = 0 \Rightarrow 10 - 4t = 0$

$\Rightarrow t = 2.5$

Using $\quad s = ut + \frac{1}{2}at^2$

when $t = 2.5$ $\quad s = 10 \times 2.5 - 2 \times 6.25$

$s = 12.5$

when $t = 4$ $\quad s = 10 \times 4 - 2 \times 16$

$s = 8.$

The total distance travelled between $t = 0$ and $t = 4$ is $12.5 \text{ m} + 4.5 \text{ m} = 17 \text{ m}$.

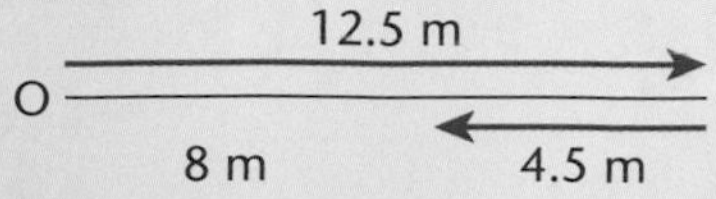

2 (a) The acceleration is positive between $t = 0$ and $t = 25$ so the greatest velocity occurs when $t = 25$.

(b)

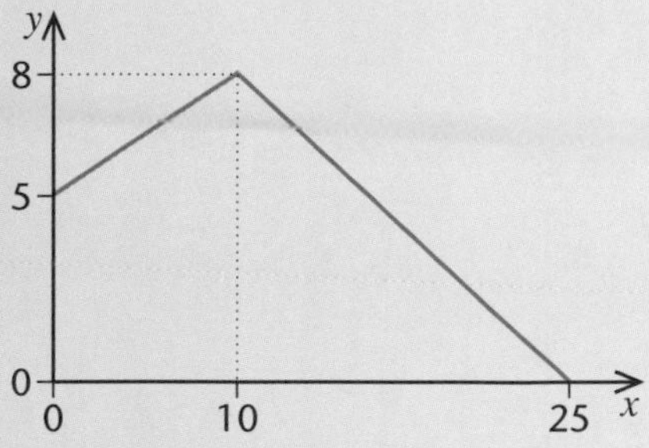

The area under the graph represents the increase in velocity between $t = 0$ and $t = 25$.

Area $= \frac{1}{2} \times 10(5 + 8) + \frac{1}{2} \times 15 \times 8 = 125.$

The increase in velocity is 125 m s^{-1}.

The initial velocity is 20 m s^{-1}.

The maximum value of the velocity is 145 m s^{-1}.

3 (a) The acceleration is constant so the constant acceleration formulae may be used.

$\mathbf{u} = 2\mathbf{i} + \mathbf{j}$

$\mathbf{a} = \mathbf{i} + 3\mathbf{k}$

$t = 4.$

Using $\quad \mathbf{s} = \mathbf{u}t + \frac{1}{2}\mathbf{a}t^2$

$\mathbf{s} = (2\mathbf{i} + \mathbf{j})4 + (\mathbf{i} + 3\mathbf{k})8$

$\mathbf{s} = 16\mathbf{i} + 4\mathbf{j} + 24\mathbf{k}.$

This is the displacement from P to Q

so $\quad \overrightarrow{PQ} = 16\mathbf{i} + 4\mathbf{j} + 24\mathbf{k}.$

(b) $\overrightarrow{OQ} = \overrightarrow{OP} + \overrightarrow{PQ}$

$= -5\mathbf{i} + 11\mathbf{j} + \mathbf{k} + 16\mathbf{i} + 4\mathbf{j} + 24\mathbf{k}$

$= 11\mathbf{i} + 15\mathbf{j} + 25\mathbf{k}.$

(c) Average speed $= \dfrac{\text{distance}}{\text{time}} = \dfrac{|\overrightarrow{PQ}|}{4} \text{ m s}^{-1}$

$|\overrightarrow{PQ}| = \sqrt{16^2 + 4^2 + 24^2} = 29.12$

Average speed $= 7.28 \text{ m s}^{-1}$ to 2 d.p.

4

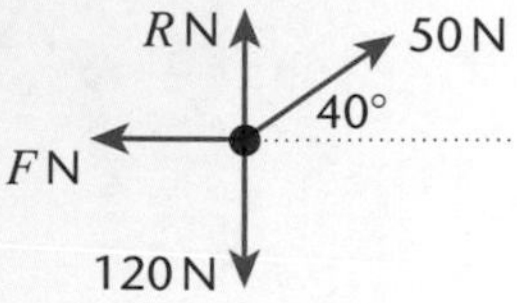

(a) Resolving vertically

$R + 50 \sin 40° - 120 = 0$

$\Rightarrow R = 120 - 50 \sin 40°$

$\Rightarrow R = 87.86 \ldots.$

The normal reaction is 87.9 N to 1 d.p.

(b) Resolving horizontally

$F - 50 \cos 40° = 0$

$\Rightarrow F = 38.30 \ldots.$

For limiting equilibrium

$F = \mu R$

giving $\quad 38.30\ldots = \mu \times 87.86\ldots$

$\mu = 0.4359 \ldots.$

The coefficient of friction is 0.436 to 3 d.p.

5

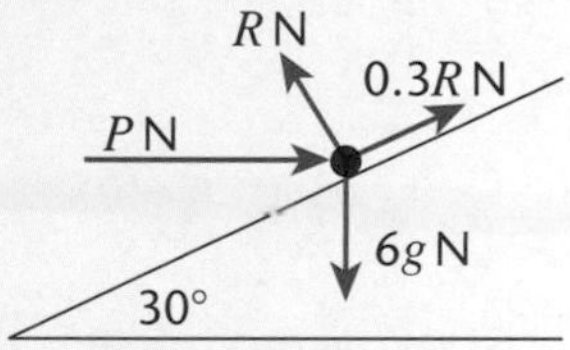

The equations are simpler to deal with by resolving vertically and horizontally, in this case, instead of parallel and normal to the plane.

The minimum value of P corresponds to the case where friction is limiting, as shown in the diagram.

Resolving vertically

$R \cos 30° + 0.3 R \sin 30° = 6 \times 9.8$

$$R = \frac{6 \times 9.8}{\cos 30° + 0.3 \sin 30°}$$

$R = 57.87 \ldots.$

Resolving horizontally

$P + 0.3R \cos 30° - R \sin 30° = 0$

$P = R \sin 30° - 0.3R \cos 30°$

$P = 13.90 \ldots.$

The minimum value of P is 13.9 to 1 d.p.

Mechanics 1 (continued)

6

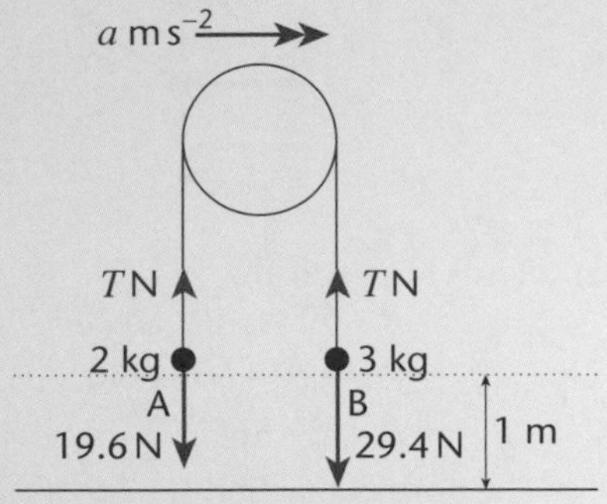

(a) The equations of motion for the particles are:

For A $\quad T - 19.6 = 2a$

For B $\quad 29.4 - T = 3a.$

Adding gives $9.8 = 5a \Rightarrow a = 1.96$

The acceleration of the system is 1.96 m s^{-2}

(b) $T - 19.6 = 3.92 \Rightarrow T = 23.52$

The tension in the string is 23.52 N.

(c) For B $\quad u = 0$

$a = 1.96,$

$s = 1.$

Using $\quad s = ut + 1/2at^2$

$1 = 0 + 0.98t^2 \ (t > 0)$

$\Rightarrow t = 1.01 \ldots .$

Particle B hits the ground after 1.01 s to 2 d.p.

(d) using $\quad v = u + at$

$v = 0 + 1.96 \times 1.01 \ldots = 1.979 \ldots .$

Particle B hits the ground with speed 1.98 m s^{-1} to 2 d.p.

(e) Particle A rises to height 2 m then the string becomes slack and it behaves as a projectile.

Using $\quad v^2 = u^2 + 2as$

at max height $\quad 0 = 1.979^2 - 19.6s$

$\Rightarrow s = 0.2$

The max height reached by particle A is 2.2 m.

(f) In the extreme situation where the pulley does not move, each particle would be held in equilibrium and so the tension on each side would match the weight of the corresponding particle.

In the given situation, where the pulley is not light or frictionless, it cannot be assumed that the tension on each side of the pulley is the same. This introduces an extra unknown value into the equations. For example, the tension on one side might be labelled T_1 and on the other T_2.

7 (a) Before impact

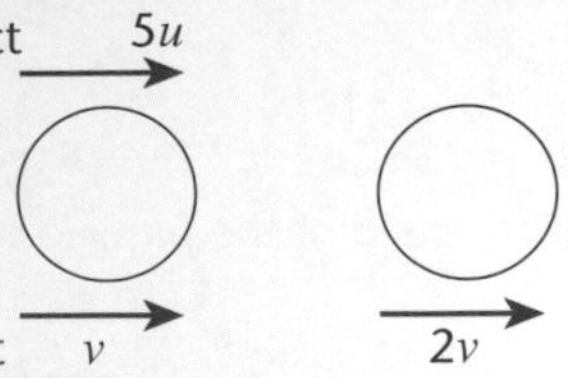

After impact

Momentum before = momentum after

so $\quad 10mu = 2mv + 2mv$

$\Rightarrow 10u = 4v$

$\Rightarrow v = 2.5u$

(b) Impulse = change in momentum

$= 5mu - 0$

$= 5mu$

Statistics 1

1 (a) $\Sigma p(x) = 1$

$\Rightarrow \frac{1}{k} + \frac{4}{k} + \frac{9}{k} = 1$

$\Rightarrow \frac{14}{k} = 1$

$k = 14.$

(b) Mean $= E(X) = \Sigma xp(x)$

$= 1 \times \frac{1}{14} + 2 \times \frac{4}{14} + 3 \times \frac{9}{14} = \frac{36}{14}$

$= \frac{18}{7}.$

$Var(X) = E(X^2) - (E(X))^2$

$= x^2 p(x) - \left(\frac{18}{7}\right)^2$

$= \frac{1}{14} + 4 \times \frac{4}{14} + 9 \times \frac{9}{14} - \left(\frac{18}{7}\right)^2$

$= \frac{19}{49}.$

(c) Mean of $3X - 5$ is $3 \times \frac{18}{7} - 5 = \frac{19}{7}.$

Variance of $3X - 5$ is $9 \times \frac{19}{49} = \frac{171}{49}.$

2 (a) $P(A \cup B) = P(A) + P(B) - P(A \cap B)$

$0.72 = 0.3 + 0.6 - P(A \cap B)$

$\Rightarrow P(A \cap B) = 0.18$

As $P(A \cap B) \neq 0$

A and B are not mutually exclusive.

(b) $P(A) \times P(B) = 0.3 \times 0.6 = 0.18$

As $P(A \cap B) = P(A) \times P(B)$,

A and B are independent.

(c) $P(A \cup B)' = 1 - P(A \cup B) = 0.28$

(d) $P(A' \cup B') = P(A') + P(B') - P(A' \cap B')$

$= 0.7 + 0.4 - 0.7 \times 0.4$

$= 0.82$

3 (a) (i) $P(RRR) = \frac{26}{52} \times \frac{25}{51} \times \frac{24}{50} = \frac{2}{17}$

(ii) P(same colour) $= P(RRR) + P(BBB)$

$= \frac{2}{17} + \frac{2}{17} = \frac{4}{17}.$

(iii) P(at least one red) $= 1 - P(BBB)$

$= 1 - \frac{2}{17} = \frac{15}{17}.$

(b) Using A to represent getting an ace and D to represent getting a diamond:

$P(A|D') = \frac{P(A \cap D')}{P(D')} = \frac{\frac{3}{52}}{\frac{3}{4}} = \frac{3}{52} \times \frac{4}{3}$

$= \frac{1}{13}.$

4 Using X to represent the lifetime of a battery in hours:

(a) $P(X > 8.5) = P\left(Z > \frac{8.5 - 8}{\frac{25}{60}}\right)$

$= P(Z > 1.2)$

$= 1 - \Phi(1.2)$

$= 1 - 0.8849\ldots$

$= 0.1151\ldots.$

So the probability that a battery selected at random will last more than $8\frac{1}{2}$ hours is 0.115 to 3 d.p.

(b) $P(X < 7) = P\left(Z < \frac{7 - 8}{\frac{25}{60}}\right)$

$= P(Z < -2.4)$

$= \Phi(-2.4)$

$= 1 - \Phi(2.4)$

$= 1 - 0.9918$

$= 0.0082$

The percentage of batteries that will last for less than 7 hours is 0.82%

Statistics 1 (continued)

5 (a) $r = \dfrac{S_{xy}}{\sqrt{S_{xx}S_{yy}}}$

$$= \frac{28785 - \dfrac{567 \times 484}{10}}{\sqrt{\left\{\left(34216 - \dfrac{576^2}{10}\right)\left(24450 - \dfrac{484^2}{10}\right)\right\}}}$$

$= 0.879$ to 3 d.p.

(b) $b = \dfrac{S_{xy}}{S_{xx}} = \dfrac{28785 - \dfrac{576 \times 484}{10}}{34216 - \dfrac{576^2}{10}}$

$= 0.87307\ldots = 0.873$ (3 d.p.)

$a = \dfrac{484}{10} - 0.873\ldots \times \dfrac{576}{10} = -1.88906\ldots$

The equation is $y = -1.889 + 0.873x$.

(c) $\hat{y} = -1.889 + 0.873 \times 65 = 55$ (2 s.f.)

If the assumption that physics scores depend on maths scores is correct, the estimate should be reliable, since $x = 65$ is within the range of data and the high value of r suggests a strong linear correlation between the variables.

Decision Mathematics 1

1 (a) Total memory to transfer = 4520 Kb = 4.52 Mb

$\dfrac{4.52}{1.4} = 3.228\ldots$

Three disks do not have sufficient capacity to hold all of the files. A minimum of four disks is required.

(b)

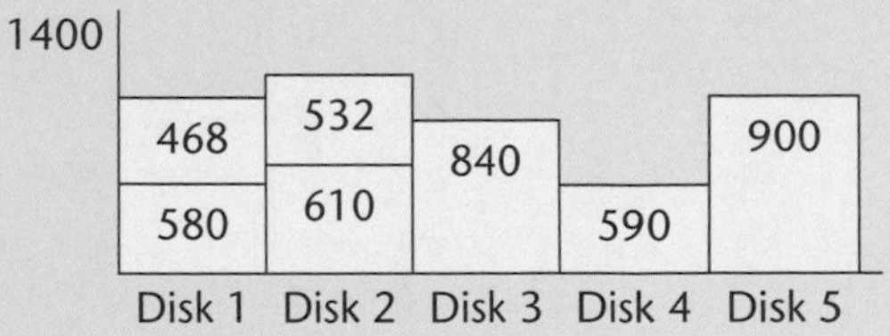

(c) Writing the memory sizes in order gives 900, 840, 610, 590, 580, 532, 468.

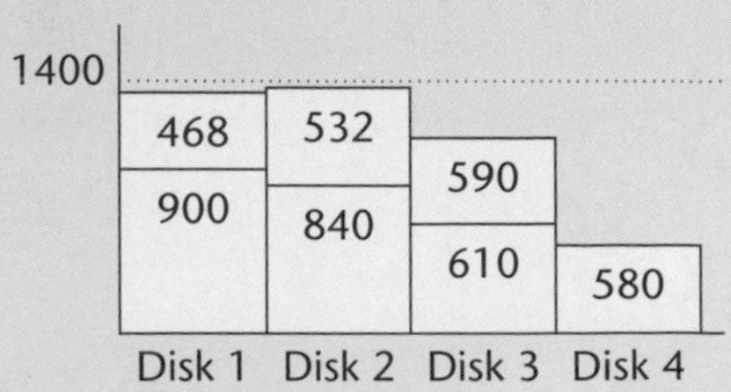

2 Applying the matrix form of Prim's algorithm gives:

	1	7	5	6	3	2	4
	A	B	C	D	E	F	G
~~A~~	–	50	40	61	–	38	–
~~B~~	50	–	–	–	–	(47)	54
~~C~~	(40)	–	–	35	–	–	–
~~D~~	61	–	(35)	–	42	–	–
~~E~~	–	–	–	42	–	(25)	10
~~F~~	(38)	47	–	–	25	–	63
~~G~~	–	54	–	–	(10)	63	–

The minimum cost is found by adding the circled figures and multiplying by £100.

Minimum cost = £19 500.

3

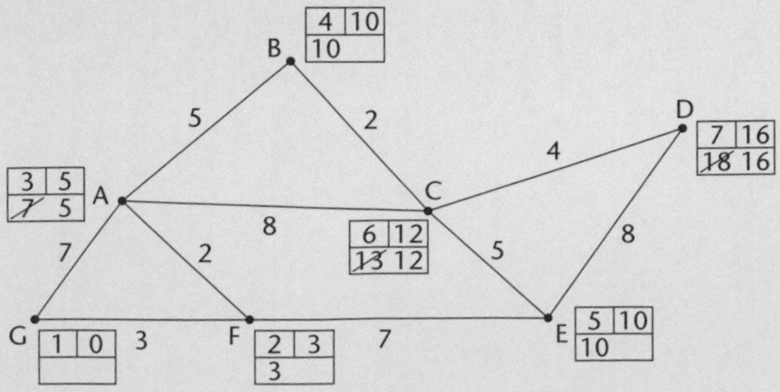

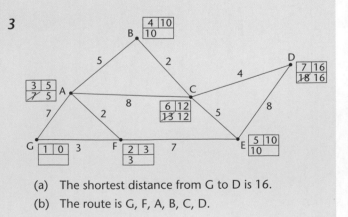

(a) The shortest distance from G to D is 16.

(b) The route is G, F, A, B, C, D.

Decision Mathematics 1 *(continued)*

4

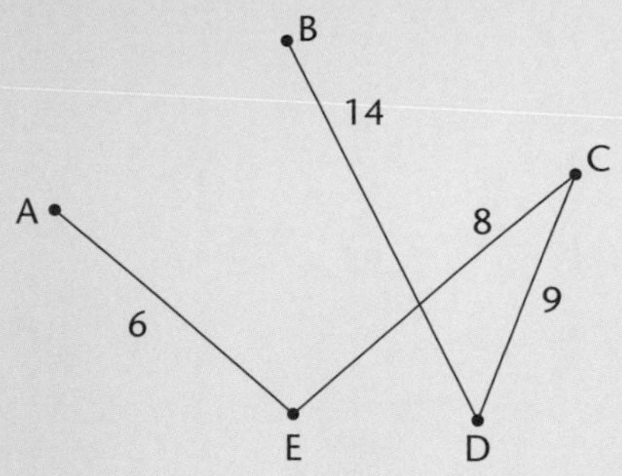

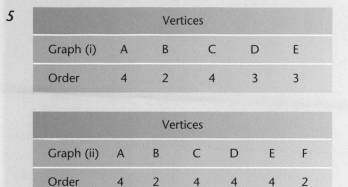

(a) The order in which the edges were connected is:
AE, EC, CD, AB (not AD or AC to avoid cycles).
The length of the minimum spanning tree is 37.

(b) Using the short-cut from B to A gives an upper bound for the TSP as $37 + 12 = 49$.

5

Vertices					
Graph (i)	A	B	C	D	E
Order	4	2	4	3	3

Vertices						
Graph (ii)	A	B	C	D	E	F
Order	4	2	4	4	4	2

Vertices				
Graph (iii)	A	C	D	E
Order	3	3	3	3

6 (a) The objective function is $4a + 9b + 12c$.

(b) The formulation of the task as a linear programming problem is:

Maximise: $P = 4a + 9b + 12c$

Subject to: $6a + 15b + 20c \leqslant 100$

$a + b + c \leqslant 10$

and $a \geqslant 0, b \geqslant 0, c \geqslant 0$.

(c) In the solution a, b and c must be integers.

7 (a)

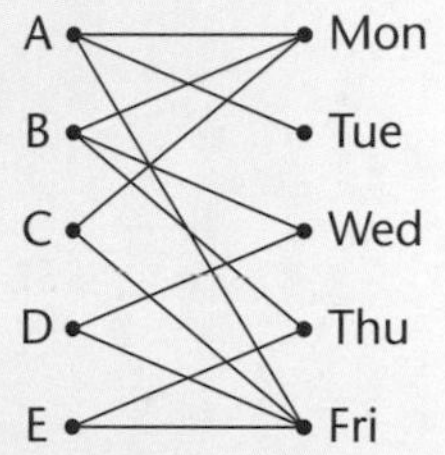

(b)

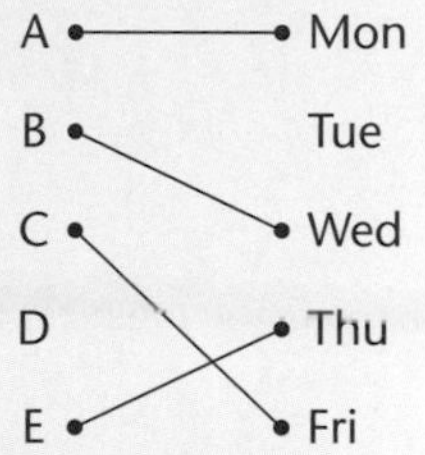

(c)

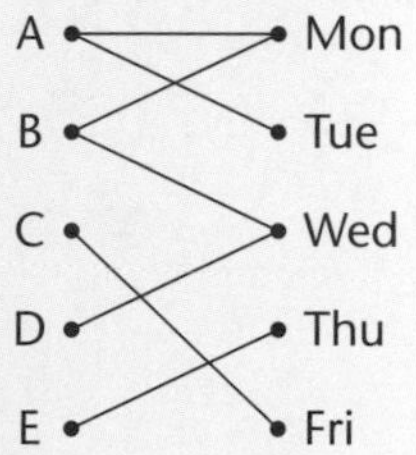

The alternating path is:

D – Wed – B – Mon – A – Tue.

The new matching is:

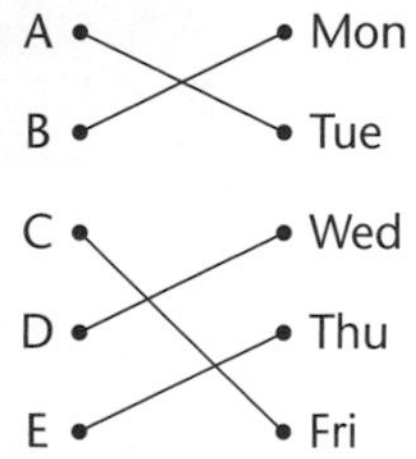

Notes

Index